P9-CLC-074

DISCARD

Chicago Public Library

REFERENCE

Form 178 rev. 11-00

Avalon Branch Library
8828 So. Stony Island Ave.
Chicago, ILL 60617

Teen Guides to

Environmental Science

Teen Guides to
Environmental Science

Earth Systems and Ecology
VOLUME I

John Mongillo

with assistance from Peter Mongillo

Greenwood Press
Westport, Connecticut • London

Library of Congress Cataloging-in-Publication Data

Mongillo, John F.
    Teen guides to environmental science / John Mongillo with assistance from Peter Mongillo.
      p. cm.
    Includes bibliographical references and index.
    Contents: v. 1. Earth systems and ecology—v. 2. Resources and energy—v. 3. People
and their environments—v. 4. Human impact on the environment—v. 5. Creating a
sustainable society.
    ISBN 0–313–32183–3 (set : alk. paper)—ISBN 0–313–32184–1 (v. 1 : alk. paper)—
ISBN 0–313–32185–X (v. 2 : alk. paper)—ISBN 0–313–32186–8 (v. 3 : alk. paper)—
ISBN 0–313–32187–6 (v. 4 : alk. paper)—ISBN 0–313–32188–4 (v. 5 : alk. paper)
    1. Environmental sciences. 2. Human ecology. 3. Nature–Effect of human beings on. I.
Mongillo, Peter A. II. Title.
    GE105.M66   2004
    333.72—dc22   2004044869

British Library Cataloguing in Publication Data is available.

Library of Congress Catalog Card Number: 2004044869
ISBN: 0–313–32183–3 (set)
        0–313–32184–1 (vol. I)
        0–313–32185–X (vol. II)
        0–313–32186–8 (vol. III)
        0–313–32187–6 (vol. IV)
        0–313–32188–4 (vol. V)

First published in 2004

Greenwood Press, 88 Post Road West, Westport, CT 06881
An imprint of Greenwood Publishing Group, Inc.
www.greenwood.com

Printed in the United States of America

The paper used in this book complies with the
Permanent Paper Standard issued by the National
Information Standards Organization (Z39.48–1984).

10 9 8 7 6 5 4 3 2 1

CONTENTS

CHAPTER 3 How Ecosystems Work   40

CHAPTER 4 Land Biomes: Forests   56

CHAPTER **5** Other Land Biomes 79

CHAPTER **6** Water Biomes 101

CHAPTER **7** Changes in the Ecosystem 123

ACKNOWLEDGMENTS

The authors wish to acknowledge and express the contribution of the many nongovernment organizations, corporations, colleges, and government agencies that provided assistance to the authors in the research for this book. The authors are grateful to the Greenwood Publishing Group for permission to excerpt text and photos from *Encyclopedia of Environmental Science*, John Mongillo and Linda Zierdt-Warshaw, and *Environmental Activists*, John Mongillo and Bibi Booth. Both books are excellent references for researching environmental topics and gathering information about environmental activists. Many thanks to those who provided special assistance in reviewing particular topics and offering comments and suggestions: Sara Jones, middle school director for La Jolla Country Day School in San Diego, California; Emily White, teacher of geography and world cultures at the 5th grade level at La Jolla Country Day School, San Diego, California; Lucinda Kramer and John Guido, middle school social studies coordinators, North Haven, Connecticut; Daniel Lanier, environmental professional, and Susan Santone, executive director of Creative Change, Ypsilanti, Michigan.

A special thank you goes to the following people and organizations that provided technical expertise and/or resources for photos and data: Neil Dahlstrom, John Deere & Company; Francine Murphy-Brillon, Slater Mill Historic Site; Lake Worth Public Library, Florida; Pacific Gas & Electric; Energetch; Environmental Justice Resource Center; NASA Johnson Space Center; Seattle Audubon Society; John Onuska, INMETCO; Cathrine Sneed, Garden Project; Denis Hayes, president, Bullitt Foundation; Ocean Robbins, Youth for Environmental Sanity; Maria Perez and Nevada Dove, Friends of McKinley; Juana Beatriz Gutiérrez, cofounder and president of Madres del Este de Los Angeles—Santa Isabel; Mikhail Davis, director, Brower Fund, Earth Island Institute; Randall Hayes, president, Rainforest Action Network; Tom Repine, West Virginia Geologic Survey; Peter Wright and Nancy Trautmann, Cornell University; Mary N. Harrison, University of Florida; and Huanmin Lu, University of Texas, El Paso.

Other sources include Centers for Disease Control and Prevention, Department of Environmental Management, Rhode Island; ChryslerDaimler; Pattonville High School; National Oceanic and

Atmospheric Administration; Chuck Meyers, Office of Surface Mining; U.S. Department of Agriculture; U.S. Fish and Wildlife Service; U.S. Department of Energy; U.S. Environmental Protection Agency; U.S. National Park Service; National Renewable Energy Laboratory; Tower Tech, Inc.; Earthday 2000; Marilyn Nemzer, Geothermal Education Office; U.S. Agricultural Research Service; U.S. Geological Survey; Glacier National Park; Monsanto; CREST Organization; Shirley Briggs, Vortec Corporation; National Interagency Fire Center/Bureau of Land Management; Susan Snyder, Marine Spill Response Corporation; Lisa Bousquet, Roger Williams Park Zoo, Rhode Island; Netzin Gerald Steklis, International National Response Corporation; U.S. Department of the Interior/Bureau of Reclamation; Bluestone Energy Services; OSG Ship Management, Inc.; and Sweetwater Technology.

In addition, the authors wish to thank Hollis Burkhart and Janet Heffernan for their copyediting and proofreading support; Muriel Cawthorn, Hollis Burkhart, and Liz Kincaid for their assistance in photo research; and illustrators Christine Murphy, Susan Stone, and Kurt Van Dexter.

The responsibility of the accuracy of the terms is solely that of the authors. If errors are noticed, please address them to the authors so that corrections can be made in future revisions.

INTRODUCTION

Teen Guides to Environmental Science is a reference tool which introduces environmental science topics to middle and high school students. The five-volume series presents environmental, social, and economic topics to assist the reader in developing an understanding of how human activity has changed and continues to change the face of the world around us.

Events affecting the environment are reported daily in magazines, newspapers, periodicals, newsletters, radio, and television, and on Websites. Each day there are environmental reports about collapsing fish stocks, massive wastes of natural resources and energy, soil erosion, deteriorating rangelands, loss of forests, and air and water pollution. At times, the degradation of the environment has led to issues of poverty, malnutrition, disease, and social and economic inequalities throughout the world. Human demands on the natural environment are placing more and more pressure on Earth's ecosystems and its natural resources.

The challenge in this century will be to reverse the exploitation of Earth's resources and to improve social and economic systems. Meeting these goals will require the participation and commitment of businesses, government agencies, nongovernment organizations, and individuals. The major task will be to begin a long-term environmental strategy that will ensure a more sustainable society.

CREATING A SUSTAINABLE SOCIETY

Sustainable development is a strategy that meets the needs of the present without compromising the ability of future generations to meet their own needs. Many experts believe that for too long, social, economic, and environmental issues were addressed separately without regard to each other. In creating a sustainable society, there needs to be an integration of goals related to economic growth, environmental protection, and social equity. Some of these integrated sustainable goals include the following:

- Improve the quality of human life

- Conserve Earth's diversity

- Minimize the depletion of nonrenewable resources

- Keep within Earth's carrying capacity

- Enable communities to care for their own environments

- Integrate the environment, economy, and human health into decision making

- Promote caretakers of Earth.

OVERVIEW

Teen Guides to Environmental Science provides an excellent opportunity for students to study and focus on the integration of ecological, economical, and social goals in creating a sustainable society. Within the five-volume series, students can research topics from a long list of contemporary environmental issues ranging from alternative fuels and acid rain to wetlands and zoos. Strategies and solutions to solve environmental issues are presented, too. Such topics include soil conservation programs, alternative energy sources, international laws to preserve wildlife, recycling and source reduction in the production of goods, and legislation to reduce air and water pollution, just to name a few.

Major Highlights

- Assists students in developing an understanding of their global environment and how the human population and its technologies have affected Earth and its ecology.

- Provides an interdisciplinary perspective that includes ecology, geography, biology, human culture, geology, physics, chemistry, history, and economics.

- "Raises a student's awareness of a strategy called sustainable development that meets the needs of the present without compromising the ability of future generations to meet their own needs" (Bruntland Commission). The strategy includes a level of economic development that can be sustained in the future while protecting and conserving natural resources with minimum damage to the environment. People concerned about sustainable development suggest that meeting the needs of the future depends on how well we balance social, economic, and environmental objectives—or needs—when making decisions today.

- Presents current environment, social, and economic issues and solutions for preserving wildlife species, rebuilding fish stocks, designing strategies to control sprawl and traffic congestion, and developing hydrogen fuel cells as a future energy source.

- Challenges everyone to become more active in their home, community, and school in addressing environmental problems and discussing strategies to solve them.

ORGANIZATION

Teen Guides to Environmental Science is divided into five volumes.

Earth Systems and Ecology

Volume I begins the discussion of Earth as a system and focuses on ecology—the foundation of environmental science. The major chapters examine ecosystems, populations, communities, and biomes.

Resources and Energy

Currently, fossil fuels drive the economy in much of the world. In Volume II conventional fuels such as petroleum, coal, and natural gas are reported. Other chapters elaborate on nuclear energy, hydrogen energy, wind energy, geothermal energy, solar energy, and natural resources such as soil and minerals, forests, water resources, and wildlife preserves.

People and Their Environments

The history of civilizations, human ecology, and how early and modern societies have interacted with the environment is presented in Volume III. The major chapters highlight the Agricultural Revolution, the Industrial Revolution, global populations, and economic and social systems.

Human Impact on the Environment

Volume IV discusses the causes and the harmful effects of air and water pollution and sustainable solution strategies to control the problems. Other chapters examine the human impact on natural resources and wildlife and discuss efforts to preserve them.

Creating a Sustainable Society

Volume V focuses on the importance of living in a sustainable society in which generations after generations do not deplete the natural resources or produce excessive pollutants. The chapters present an overview of sustainability in producing products, preserving wildlife habitats, developing sustainable communities and transportation systems, and encouraging sustainable management practices in agriculture and commercial fishing. The last chapter in this volume considers the importance of individual activism in identifying and solving environmental problems in one's community.

PROGRAM RESEARCH

The five-volume series represents research from a variety of recurring and up-to-date sources, including newspapers, middle school and high school textbooks, trade books, television reports, professional journals, national and international government organizations, nonprofit organizations, private companies, businesses, and individual contacts.

CONTENT STANDARDS

The series provides a close alignment with the fundamental principles developed and reported in the President's Council on Sustainable Development and the learning outcomes for middle school education standards found in the North American Association for Environmental Education, the National Geography Standards, and the National Science Education Standards.

MAJOR ENVIRONMENTAL TOPICS

The *Teen Guides to Environmental Science* provide terms, topics, and subjects covered in most middle school and high schools environmental science courses. These major topics of environmental science include, but are not limited to:

- Agriculture, crop production, and pest control
- Atmosphere and air pollution
- Ecological economies
- Ecology and ecosystems
- Endangered and threatened wildlife species
- Energy and mineral resources
- Environmental laws, regulations, and ethics
- Oceans and wetlands
- Nonhazardous and hazardous wastes
- Water resources and pollution.

SPECIAL FEATURES

Tables, Figures, and Maps

Hundreds of photos, tables, maps, and figures are ideal visual learning strategies used to enhance the text and provide additional information to the reader.

Vocabulary

The vocabulary at the end of each chapter provides a definition for a term used within the chapter with which a reader might be unfamiliar.

Marginal Topics

Each chapter contains marginal features which supplement and enrich the main topic covered in the chapter.

Activities

More than 100 suggested student research activities appear at the ends of the chapters in the books.

In-Text References

Many of the chapters have specially marked callouts within the text which refer the reader to other books in the series for additional information. For example, fossil fuels are discussed in Volume V; however, an in-text reference refers the reader to Volume II for more information about the topic.

Websites

A listing of Websites of government and nongovernment organizations is available at the end of each chapter allowing students to research topics on the Internet.

Bibliography

Book titles and articles relating to the subject area of each chapter are presented at the end of each chapter for additional research opportunities.

Appendixes

Four appendixes are included at the end of each volume:

- Environmental Timeline, 1620–2004. To understand the history of the environmental movement, each book provides a comprehensive timeline that presents a general overview of activists, important laws and regulations, special events, and other environmental highlights over a period of more than 400 years.

- Endangered List of U.S. Wildlife Species by State.

- Website addresses by classification.

- Government and nongovernment environmental organizations.

CHAPTER 1

Earth Systems

Earth is the only place we know of that can support life. (Courtesy of NASA/Johnson Space Center)

The photograph of Earth's surface shown on this page was produced by a Landsat satellite. The satellite scans the entire surface of Earth in just 16 days. Special equipment aboard the satellite records digital images which show landforms such as rivers, mountains, farmlands, and forests. Images from other satellites have transmitted weather conditions in the atmosphere, the migration of land animals, and the flow of ocean currents to Earth stations. These thousands of images help confirm that Earth is a complex planet made up of several parts—land, water, air, and living things—all of which make up our *environment* on Earth.

For convenience, scientists learn about Earth by observing and studying its smaller portions or units. These units can be referred to as *systems*. The major systems of Earth include the lithosphere, the atmosphere, the hydrosphere, and the biosphere. Knowing how each of these systems interacts with each of the others helps us understand how our Earth functions as a system.

BIOSPHERE

The portion of Earth that includes all living organisms and the parts of Earth in which they live is known as the biosphere. The word biosphere comes from the Greek word bios, meaning "life," and the Latin word sphere, which means "total range." Thus, the biosphere contains every living species on Earth, including human beings.

The biosphere includes the portions of Earth's atmosphere, lithosphere, and hydrosphere that are suitable for sustaining life. Living organisms need special conditions which exist in the biosphere to support life. These conditions typically include a supply of water, solar energy, suitable climate, chemical elements and minerals, and air.

Compared to the total size of Earth, the biosphere is quite small. Living things may be found up to about 8 kilometers (5 miles) from Earth's surface, on and just below the surface, and within bodies of water. Plants exist at 300 meters (900 feet) below sea level and at 6,100 meters (18,500 feet) above sea level. Evidence exists for the presence of bacteria in petroleum deposits at depths of about 2,000 meters (7,000 feet) below the surface. Other kinds of bacteria have been found at 9,750 meters (30,000 feet) above sea level. Some kinds of spiders live at 6,700 meters (21,000 feet) above sea level. However, most living things live within the upper 100 meters (300 feet) of the lithosphere and hydrosphere.

In the late 1960s, James Lovelock presented the Gaia hypothesis which states that Earth is not an inanimate habitat. Gaia is the name of

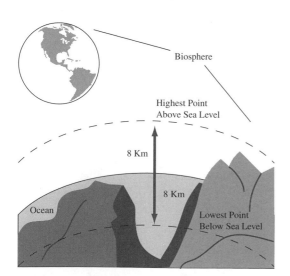
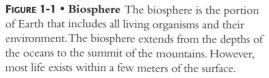

FIGURE 1-1 • Biosphere The biosphere is the portion of Earth that includes all living organisms and their environment. The biosphere extends from the depths of the oceans to the summit of the mountains. However, most life exists within a few meters of the surface.

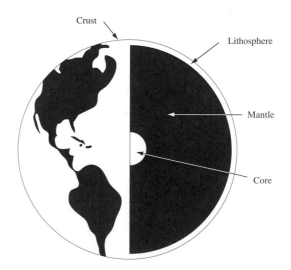

FIGURE 1-2 • Diagram of the Lithosphere The lithosphere is the solid outer shell of Earth. It is composed of the crust and the uppermost part of the mantle. The two kinds of crust include the continental crust and the oceanic crust. Much of Earth's surface is covered by water in the oceans and seas.

a Greek Earth goddess. Lovelock wrote that Earth's surface behaves as an organism capable of controlling its own composition and its environment. Although Lovelock's hypothesis falls outside of the realm of science, his idea does encourage humans to care for and respect planet Earth and its systems as they would care for all living organisms.

LITHOSPHERE

The materials of Earth occur in the form of solid, liquids, or gases. The solid materials, or the rocky, outer layer of Earth, is called the lithosphere. The lithosphere is made up of all of Earth's land areas, including those lying beneath the oceans. The land areas make up about 30 percent of the surface of Earth. The remaining areas are below sea level.

Earth's Crust and Mantle

The lithosphere, which is composed of Earth's thin outer shell, includes the crust and the cooler, solid upper mantle. On average, the lithosphere is about 100 kilometers (62 miles) thick. The crust, including the continents and the ocean-floor crust, is composed mainly of rocks such as granite and basalt.

Rocks

The major classes of rocks—igneous, sedimentary, and metamorphic—are based on their origins. Igneous rocks are formed by the cooling and hardening of hot liquid rock beneath Earth's surface, called magma. Examples of igneous rocks include basalt, granite, and gabbro. Sedimentary rocks are formed from loose rock materials called sediments and the remains of plants and animals that lived in the shallow oceans millions of years ago. Over time, these materials were cemented together under chemical action and pressure. Examples of sedimentary rocks include sandstone, shale, and limestone. Metamorphic rocks were formed by the effect of heat and pressure on other rocks. Examples include slate, marble, and gneiss. To date, the oldest rocks found on Earth are the igneous rocks that were discovered in northwest Canada, which are believed to be about 3.96 billion years old.

Rocks are made up of minerals, which are groups of chemical compounds. For example, some granites are composed of the minerals quartz, feldspar, and mica. Minerals also include ores such as hematite, an iron ore, and bauxite, an aluminum ore. Gems are another type of mineral. These valuable minerals include diamonds, emeralds, and rubies.

Earth's Surface Is Broken Up

The surface layer of the lithosphere is not made up of one continuous piece like the covering of a ball. According to the theory of plate tectonics, the lithosphere is actually broken up into a number of enormous blocks or "plates" of different shapes and sizes. The plates, which fit together like pieces of a jigsaw puzzle, move relatively or closely to

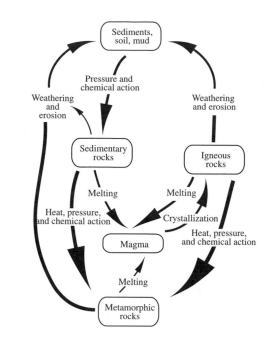

FIGURE 1-3 • Rock Cycle
A rock cycle shows how rocks change into other kinds of rocks from one kind to another. The changes are due to heat, pressure, melting, weathering, chemical action, erosion, and climate conditions.

| TABLE 1-1 | Common Rocks |

Igneous	Sedimentary	Metamorphic
Granite	Conglomerate	Slate
Diabase	Breccia	Schist
Basalt	Sandstone	Gneiss
Obsidian	Shale, clystone	Quartzite
Pumice	Limestone Chert, flint	Marble

Note: Rocks are aggregates of a mineral or minerals, which are inorganic materials with a definite structure and physical and chemical properties.

each other. Each plate is about 100 kilometers (60 miles) thick and moves from about 2.5 to 15 centimeters (about 1 to 6 inches) a year. The major plates are known as the North American Plate, the South American Plate, the Eurasian Plate, the Indian–Australian Plate, the African Plate, the Pacific Plate, and the Antarctic Plate. There are also several smaller plates, such as the Caribbean Plate, which fit in between the major plates.

Crustal plates are created, destroyed, or move past each other in a variety of ways. In some areas, such as the Mid-Atlantic, the plates move apart or diverge. In other places, they move close together or converge. In some areas where their edges meet, the plates grind together as they slide past or dive beneath one another. When the plates of the lithosphere rub and grind against one another, energy builds up along plate boundaries. It is here that most mountains form, volcanoes erupt, and earthquakes occur. For example, the movement of the Pacific Plate against the North American Plate has resulted in earthquake activity in

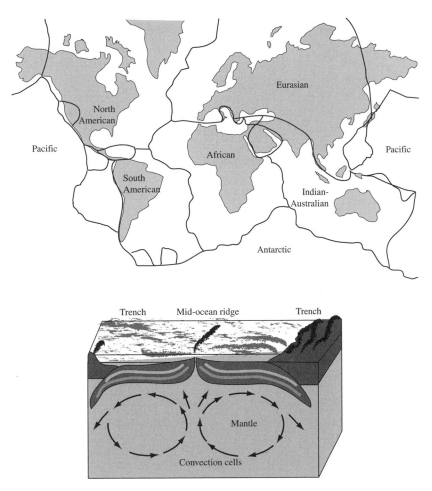

FIGURE 1-4 • **Tectonic Plates** Earth's lithosphere, the solid outer layer, is covered by a number of plates of different shapes and sizes that move relative to each other. Some of the major plates include North American, South American, Eurasian, Indian-Australian, African, Pacific, and Antarctic plates.

FIGURE 1-5 • **Convection Currents** According to one hypothesis of plate motions, convection currents in the mantle are thought to be the mechanism that causes the plates to move.

California. Some mountain ranges, such as the Himalayas in Asia, were formed by the collision of the Indian-Australian Plate and the Eurasian Plate. Even today, the Himalayas are still rising above sea level because of the collision. The highest mountain in the Himalayas is Mount Everest, which is approximately 8,846 meters (29,022 feet) high.

According to *geologists*, convection currents in the mantle is thought to be the mechanism that moves the plates. Convection creates thermal currents within Earth's mantle. These thermal currents rise to the surface and then sink, as a result of heating and cooling. When convection currents rise, the affected plates move apart and spread out, generating new crust. When convection currents sink, the activity causes the plates to come together at their boundaries. In this process, part of Earth's crust is destroyed as the edge of one plate "dives" under another neighboring plate. Although the general theory of plate tectonics is now widely accepted by scientists, many aspects of the theory require further investigation.

Plate movement is of environmental importance because tectonic activity has greatly affected Earth's surface and its living organisms for more than millions, perhaps billions, of years.

DID YOU KNOW?

A tsunami is a very large ocean wave caused by underwater earthquakes or volcanic eruptions.

Continental Drift

In 1912 Alfred Wegener (1880–1930), a German meteorologist, *hypothesized* that all of Earth's masses had once been joined together as one large landmass called *Pangaea*. The hypothesis further states that Pangaea broke apart about 200 million years ago, and that the continents have gradually drifted into the positions they occupy today. This is known as continental drift. Wegener's hypothesis was the first detailed study made on continental drift. He incorporated evidence from rocks and fossils from different parts of the world to show that distant continents were once joined. Wegener's hypothesis was rejected by many scientists of his generation because he was not able to explain how or why the drifting occurred.

During the 1960s through the 1980s, the basic concepts and hypotheses of continental drift became generally accepted by earth scientists. Theories of continental drift, plate tectonics, and sea-floor spreading developed significantly during this period, marking a revolution in earth science. According to the theory of plate tectonics, Earth's crust consists of at least 15 sections, or plates, which "float" on Earth's mantle, the semisolid layer just below the crust. As the plates move slowly over time, the positions of the continents and the ocean basins change, and this movement continues today.

FIGURE 1-6 • Pangaea
Pangaea as it looked about 200 million years ago. The continents were all joined into one supercontinent; however, at that time, the large landmass began to break apart into the present continent landforms.

WEATHERING AND SOIL FORMATION

Earth's crust is modified by the processes of *weathering* that occur at or near the surface. Weathering caused by precipitation, wind, glaciers, running water, and other natural events, such as landslides, can gradually break down rock.

Physical weathering is the breaking up of rocks into smaller pieces. Water gets into the cracks of the rock and then it freezes. The freezing water exerts pressure on the rock as it expands and contracts, thereby breaking the rock apart. Rocks can also be worn down and broken up by particles carried by strong winds and fast moving rivers and streams. People can break up rocks when building highways, dams, and other structures. The growing roots of trees and other plants can also apply pressure on the surrounding rocks.

The release of chemical substances by organisms into the environment can lead to the chemical weathering of rock. For example, lichens often live on bare rock. As they carry out their life processes, these organisms give off acids that can slowly break down the rock. In time, the weathered broken rock can become *soil*.

FIGURE 1-7 • Soil Composition The formation of soil begins when rocks break down and weather into smaller fragments. The process can take thousands of years. However, in time, plants begin to grow in the weathered rock. Later on, fungi, algae, insects, worms, snails, and other animals add organic matter to the soil. Earthworms can move 20 tons of soil per acre every year. A dark rich humus material is formed when the plants and animals in the soil die and decay. The humus provides fertile soil and helps retain water.

Soil Development

Soil is an important natural resource because it is essential to living organisms on Earth. Weathering is the first step in the formation of soil because it breaks down the solid bedrock into small particles. During the weathering process, chunks of rock continue to break up into smaller and smaller pieces. In time, many kinds of organisms, such as bacteria, fungi, and plants begin to live in the weathered materials. Over time, the organisms die and decay and add organic matter and nutrients to the weathered materials forming soil. Climate, *topography*, and the daily activities of plants and animals also play important roles in the development of soils.

ATMOSPHERE

Atmospheric Gases

The protective layer of gases that surrounds Earth is known as the atmosphere. Much of the mass of the gaseous atmosphere is within 32 kilometers (20 miles) of Earth's surface. The atmosphere is made up of about 78 percent nitrogen and 21 percent oxygen. Oxygen, an odorless, colorless, and tasteless gas, is essential for most living organisms. Oxygen is made up of two atoms (O_2) in the atmosphere and three atoms (O_3) in the ozone layer. Oxygen is produced by plants, cyanobacteria, algae, and

Human Activities Can Speed Up the Weathering Process

Not all of the weathering that occurs in the environment results from natural phenomena. For example, pollutants released into the environment through activities carried out by humans can speed or promote weathering. Nitrogen oxides (NO_x) and sulfur dioxide (SO_2) are pollutants. They are released into the environment as a result of combustion, the burning of fossil fuels. Once in the atmosphere, these compounds can combine with water vapor to form acids which are carried back to Earth's surface through precipitation such as rain, snow, and fog. Acid precipitation that falls into lakes or onto forest soils can be extremely harmful to the organisms living in these ecosystems. The acid rain that falls onto structures on Earth's surface, such as buildings, monuments, and metal structures, can break down these substances, causing them to fall apart.

Refer to Volume II for more information about soil and minerals.

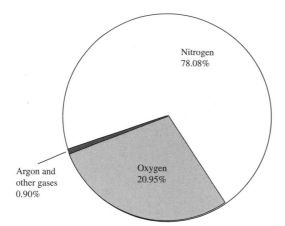

FIGURE 1-8 • Composition of Air The air surrounding Earth is composed of mostly nitrogen and oxygen gases. Some of the other gases include carbon dioxide, argon, and helium. Most of the gases in the air are colorless and have no smell.

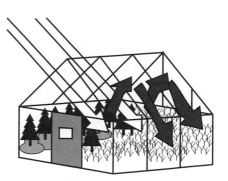

FIGURE 1-9 • Greenhouse Effect The atmosphere works like a greenhouse trapping the Sun's heat and warming Earth. The glasswalls of the greenhouse, acting much like the atmosphere, traps the sun's heat from escaping back into space.

protists. The oxygen (O_2) is released to the environment as a byproduct of *photosynthesis*. Nitrogen is one of the most abundant, naturally occurring elements on Earth. It is important for the growth, reproduction, and metabolism of living organisms. Other trace gases in the atmosphere include water vapor, argon, carbon dioxide, neon, helium, methane, hydrogen, ammonia, carbon monoxide, and ozone. The atmosphere also contains particulate matter (dust) and aerosols (mists).

Greenhouse Effect

Certain atmospheric gases help keep Earth warm and habitable. Like the glass panes of a greenhouse, certain atmospheric gases, such as carbon dioxide and methane, keep heat in and make Earth hospitable to organisms. The gases allow solar energy, or energy from the sun, to pass through the atmosphere and be absorbed at Earth's surface. But they also trap most of the radiant heat emitted from Earth's surface in the lower atmosphere, preventing it from escaping back into space. Energy resulting from the greenhouse effect influences the heating of the ground surface, the melting of ice and snow, the evaporation of water, and photosynthesis in plants. Without the greenhouse effect, Earth's average global temperature would be −18°C (−0.4°F), rather than its current 15°C (59°F). Even the oceans would be frozen under such harsh conditions. Life would not exist.

Layers of the Atmosphere

Earth's atmosphere is divided into several layers. Most scientists, however are concerned mainly with two layers—the troposphere and the stratosphere. The troposphere is the region of the atmosphere that is

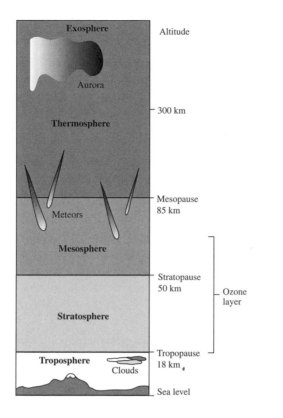

FIGURE 1-10 • Layers of the Atmosphere The atmosphere is the protective layer of gases that surround Earth and plays a key role in Earth's ecosystem and biogeochemical cycles. The major gases are oxygen and nitrogen.

closest to Earth. It includes water vapor and clouds and is a source of Earth's weather and climate. Unfortunately, it is also the place where most of Earth's pollution is found.

The troposphere extends from Earth's surface to an altitude of about 18 kilometers (11 miles), although this height varies with latitude. Temperatures decrease with altitude in the troposphere. As warm air rises, it cools, falling back to Earth, a process known as convection. Huge air movements mix the gases in the troposphere very efficiently. Chemicals, particulate matter, and other substances carried in the gases of the troposphere can be washed back to Earth by precipitation.

The stratosphere is a region that extends from about 18 kilometers to about 50 kilometers (11 to 30 miles) above Earth. Commercial aircraft fly in the lower stratosphere where there are strong steady winds and little water vapor. Warm air remains in the upper stratosphere and cool air remains in the lower stratosphere, so there is much less mixing of gases in this region than in the troposphere. As a result, convection does not occur in the stratosphere. The temperatures in the stratosphere increase with altitude and are warm due to a protective ozone layer that shields Earth

Ground-level ozone is a pollutant that can be harmful to the human respiratory system. About 10 percent of atmospheric ozone is located in the troposphere, the layer of the atmosphere nearest Earth. Much of the ozone in the troposphere forms when nitrogen oxides (NO$_x$), given off in emissions from the burning of fossil fuels or petrochemical products, react with sunlight. This ground-level ozone is a major component of photochemical smog and is one of the pollutants for which the U.S. Environmental Protection Agency (EPA) sets standards under the Clean Air Act (CAA) of 1970, as amended. Warm temperatures and stagnant high-pressure weather systems with low wind speeds contribute to harmful ozone accumulation.

Refer to Volume IV for more information about air pollution.

from much of the ultraviolet (UV) radiation, especially ultraviolet B (UV-B), given off by the sun. Such radiation is harmful to the cells and tissues of organisms. Nearly 90 percent of the Earth's ozone is found in the stratosphere.

The outer layers of the atmosphere include the mesosphere, the thermosphere, and the exosphere. The mesosphere begins at about 50 kilometers (30 miles) above Earth and extends to 85 kilometers (51 miles). In this layer there is a drop in temperature, little ozone, and very little oxygen and nitrogen. Most meteors, or rock fragments traveling in space, burn out in this layer. Above the mesophere is the thermosphere at an altitude of about 85 kilometers. Temperatures get warmer and auroras occur in this layer. Auroras or Northern Lights are streamers of different kinds of light which are visible at times around the north and south magnetic poles. Above the thermosphere is the exosphere which has no gases.

Climate

All weather conditions such as temperature, precipitation, humidity, air pressure, and winds occur in the troposphere. The average *weather* conditions occurring in an area over a long period of time is known as climate. Climate is important to all life on Earth because it determines the types of organisms that can survive in a particular area.

Many factors including latitude, air circulation, ocean currents, and the local geography help regulate and determine a region's climate. Latitude, how far north or south of the equator a place is located, is perhaps the most important factor because it has the most direct influence on average yearly temperatures.

Meteorologists recognize three general climate zones based on how much sunlight is received: tropical, polar, and temperate. At the equator

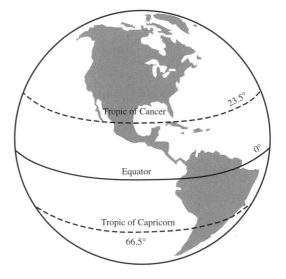

FIGURE 1-11 • Earth's Latitude Latitude is distance in degrees north and south of the equator. The tropical areas are between 23.5° north latitude and 23.5° south latitude. The polar zones are between 66.5° north and south latitudes and the poles at 90°.

(0° latitude), for example, the sun is directly overhead. Here, the sun's rays shine directly on Earth's surface. This is why tropical areas, between 23.5° north latitude and 23.5° south latitude, consistently have the hottest year-round temperatures. In polar zones (between 66.5° north and south latitudes and 90° north and south poles) the sun is lower in the sky. Therefore sunlight strikes Earth at a more oblique angle, causing it to spread out over a much larger area. Some sunlight is also lost because it is reflected by polar ice, a process known as the "albedo effect." These factors explain why polar regions have the lowest annual temperatures. Between the tropical and polar zones are the temperate zones. The continental United States, for example, is in a temperate zone. In temperate zones, weather changes with the seasons: winters are cold and summers are hot. The spring and fall seasons are characterized by mild temperatures.

The warming and cooling of air in the latitude of the tropics also helps determine the amounts of rain that fall in other parts of the world. When warm air rises, it forces the cooler upper air in the atmosphere to move toward the poles while surface air moves toward the equator. Eventually, the cooler air falls back to Earth at a latitude of about 30° north and south, warming as it falls. As this warm, dry air moves across the surface of the Earth, it creates extremely dry conditions. This is one reason why many of the world's deserts, including those in the southwestern United States, occur near 30° latitude.

Large cities can also affect local climate. When sunlight strikes areas of vegetation, much of the energy is used in evaporating moisture. In cities, on the other hand, sunlight is absorbed by streets, buildings, parking lots, cars, trucks, and buses. These objects then radiate heat back into the atmosphere. Exhaust from automobiles and other motor vehicles and emissions from factories and power plants trap this heat, creating a heat island effect. Summer temperatures in a city, for instance, can sometimes be 10°C (18°F) higher than in surrounding rural areas.

HYDROSPHERE

The hydrosphere is the part of Earth's surface that contains or is covered by water, including wetlands, ocean, rivers, streams, lakes, ponds, seas, groundwater, and atmospheric water vapor. The hydrosphere includes portions of the atmosphere and the lithosphere.

Oceans

Life has flourished on Earth for millions of years because of oceans. The world's oceans and seas make up about 71 percent of Earth's surface. Ocean water is a solution of about 3.5 percent dissolved salts. The most abundant salt in the ocean is sodium chloride, commonly known as table salt. However, table salt is only one kind of salt. Other salts present in ocean water include magnesium and sulfate. The four major oceans include the Atlantic, the Pacific, the Indian, and the Arctic. The Pacific is the largest and deepest and covers more than one-third of Earth's surface. The Arctic is largely covered by ice. Bays, gulfs, and seas are smaller bodies of the ocean that are near land or located between

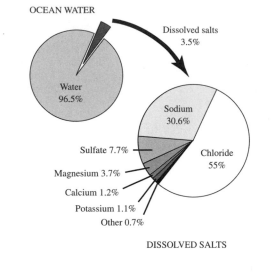

**FIGURE 1-12 •
Composition of Ocean
Salts**

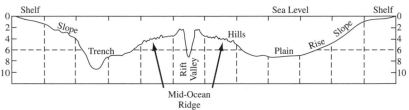

FIGURE 1-13 • Sea Floor with Mountains The diagram shows a cross section of the landscape under the ocean surface. Oceans and seas make up about 70 percent of Earth's surface. The ocean is also a huge storehouse for large quantities of carbon dioxide withdrawn from the atmosphere.

islands and land. The largest seas include the South China Sea, the Caribbean Sea, the Mediterranean Sea, and the Bering Sea.

According to data collected and analyzed by *oceanographers*, the landscape under the ocean's surface includes mountain ranges, deep trenches, undersea volcanoes, and vast flat areas or plains. Many of the undersea mountains are higher than those on land. The deepest trench in the Pacific Ocean is 11 kilometers (7 miles) deep.

OCEAN CURRENTS AND CLIMATE

Ocean currents have a major influence on global climate. The main type of ocean currents are surface currents caused by wind. The wind drives the ocean water in huge circular patterns all around the world. The currents flow in various directions because of the way in which Earth rotates on its axis. Surface currents north of the equator move to the right. Currents south of the equator are deflected to the left.

Currents distribute cold masses of water from polar regions and warm water from equatorial regions to other areas of Earth, thus influencing global climate. Because water heats up and cools down more slowly than land, many coastal areas are warmer in the winter and

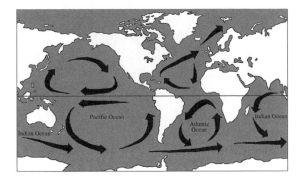

FIGURE 1-14 • Ocean Current Map Ocean surface currents are important because they help distribute heat from equatorial regions to other areas of Earth, thus influencing global weather and climate. The surface currents are driven by Earth's global wind systems.

cooler in the summer than inland areas of similar latitude. Iceland, for example, is located far north in the Atlantic Ocean. But since the Gulf Stream, a warm ocean current that flows north along the eastern edge of North America, flows past Iceland, the country has a surprisingly mild climate. Winds blowing from the ocean contain more moisture than those blowing from land. Thus, coastal regions tend to have wetter climates than places inland, particularly where the winds blow onto the coast.

Freshwater Bodies

Freshwater bodies of surface water include rivers, ponds, lakes, streams, reservoirs, wetlands, and even glaciers and ice sheets. Many major cities are located on rivers because they are a source of water and transportation. The beginning of a river occurs at high elevations. Precipitation in the form of rain or snow causes water to run down the steep slopes to form several streams. The runoff from several streams comes together at the source of the river called the headwaters. As the fast-moving water runs downstream, it cuts a V-shaped channel into the river bed. Smaller streams or rivers called tributaries feed into the main river supplying more water.

The surrounding land that supplies much of the water in a river system is called a watershed or drainage basin. The pattern of most drainage basins is shaped like the branches of a tree. A high ridge separating one watershed from another is known as a divide. Streams and rivers flow in opposite directions on either side of the divide. The river, its tributaries, and divides make up the river system. Examples of river systems include the Mississippi River and the Missouri River. Eventually all major rivers empty into oceans or large lakes.

Ponds and lakes are made up of still water—water that appears to be standing on land and not moving. Ponds and lakes are formed when water collects in low areas. They can also be formed when a river is blocked up. This event can be caused naturally by an earthquake, mudslide, volcano, or by humans or by some animals, such as beavers, when they build reservoirs. Ponds and lakes also occur in

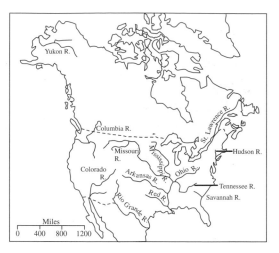

Figure 1-15 • Major Rivers of North America

low-lying areas of land. Ponds and lakes vary in size. Ponds are usually smaller than lakes and not as deep.

Wetlands

Wetlands include bogs, swamps, freshwater marshes, saltwater marshes, and swamps. A wetland is an environment that is saturated with water all or part of the year. Only a few wetlands are dry most of the year. There is a wide diversity of wetlands including prairie potholes, vernal pools, and fens as well as those already mentioned. A variety of animal and plant communities and many types of soil, topography, climate, and water content characterize each wetland. Wetlands are found on most continents even in the cold polar regions. In the United States, wetlands are found in every state, from the coastal marshes of Alaska to the mangrove forests of Florida.

DID YOU KNOW?

Wetlands stabilize shorelines thus helping to prevent erosion. Some wetlands filter out heavy metals, pesticides, and other toxic chemicals that pass through the ecosystem.

Bogs, also known as peatlands, are simply wetlands that have organic soils consisting of peat—the partially decomposed remains of plants and animals. Bogs are found in colder regions of the world where the temperatures and limited oxygen supply in the water discourage the breakdown of organic material. Fens are wetlands that are less acidic than bogs. Peat also accumulates in a fen.

Coastal saltwater marshes are open, tidal shallow wetlands which connect major water bodies with smaller lakes and streams. Coastal wetlands provide spawning and nursery grounds for a vast portion of the U.S. commercial fish and shellfish harvest.

Freshwater marshes, familiar to most Americans, make up nearly 90 percent of the wetlands in the United States. Marshes are open areas, usually with few trees and shrubs. The water in a marsh fluctuates, rising during the rainy season and disappearing during dry periods.

Prairie potholes are saucer-shaped depressions formed by retreating glaciers in the last Pleistocene glacial epoch. Prairie potholes, located in the Upper Plains states, are often called the "duck factories" of

America because of their importance to the livelihood of ducks and other migratory birds. These birds use the prairie potholes for nesting and raising their young.

Swamps are shrubby or forested wetlands, located in poorly drained areas on the edges of lakes and streams. Forest swamps exist primarily in river floodplains connected to major river systems.

Vernal pools are small, isolated wetlands that retain water on a seasonal basis, usually in the spring. The pools are vital as breeding habitats to the survival of amphibians, a class of animals that consists of frogs, toads, salamanders, and newts. Nearly 50 percent of the amphibians in the United States breed primarily in vernal pools because the pools are too shallow to support fish, the major predator to amphibian larvae, such as tadpoles.

Vocabulary

Environment Everything comprising the surroundings of an organism.

Geologists Scientists who study Earth's structure, materials, and processes.

Hypothesized Suggesting that something is true without proof.

Meteorologists Scientists who study the atmosphere. Meteorologists study wind conditions, distribution of temperature, precipitation, humidity, cloud formation and movement, thunderstorms, air pollution, and natural disasters such as hurricanes, tornadoes, and cyclones.

Oceanographers Scientists who study oceanography. Oceanography is concerned with the physical, chemical, geological, and biological properties of the environment of the oceans.

Photosynthesis The process by which light energy is absorbed and then converted to the chemical energy of glucose.

Soil The loose covering of weathered rock and decayed organic matter which can support the growth of plant life.

System An organized group of related parts that form a whole. Systems can consist, for example, of organisms, machines, galaxies, ideas, and numbers. Even your kitchen at home, the desk you work on, and the city you live in can be thought of as a system.

Topography The features of a place or region, such as mountains and valleys.

Weather The current state of the atmosphere.

Weathering The changing of rock at Earth's surface.

Activities for Students

1. Geologists believe that convection may be one cause for the movement of tectonic plates in the lithosphere. How can they use this knowledge to help predict when earthquakes will occur?

2. What factors affect the climate in your area? Spend a week tracking weather around the world for other cities at the same latitude. What similarities and differences did you notice? What may be some reasons for any differences?

3. Due to the greenhouse effect, Earth is a hospitable place suitable for human life. What could be some negative impacts of the greenhouse effect in the future, and how could we help to control them?

Books and Other Reading Materials

Anderson, Roger N. *Marine Geology*. New York: John Wiley and Sons, 1989.

Bolt, Bruce A. *Earthquakes*. New York: W. H. Freeman, 1993.

Brower, David. *Only a Little Planet*. New York: McGraw-Hill, 1972.

Cousteau, Jacques Ives. *Exploring the Wonders of the Deep*, Orlando, Fla.: Raintree/Steck-Vaughn, 1997.

Fisher, Richard V. *Volcanoes: Crucibles of Change*. Princeton, N.J.: Princeton University Press, 1998.

Graedel, Thomas E., and Paul Crutzen. *Atmosphere, Climate, and Change*. New York: W. H. Freeman, 1997.

Schaefer, Vincent J., and John A. Day. *A Field Guide to the Atmosphere*. Boston: Houghton Mifflin, 1999.

Wilson, Edward O. *Biodiversity*. Washington, D.C.: National Academy Press, 1988.

Websites

Earthguide, http://earthguide.ucsd.edu/

Landsat and Satellite Images, http://landsat.gsfc.nasa.gov/main.htm

NASA Remote Sensing Public Access Center, NASA Observatorium contains fun and games, http://www.hpcc.arc.nasa.gov/reports/annrep95/iita95/rspac_iv.htm

National Oceanographic and Atmospheric Administration, NOAA Research, http://www.oar.noaa.gov/education

National Weather Service, http://www.nws.noaa.gov

NOAA website containing the El Nino/El Nina theme page, http://www.pmel.noaa.gov/toga-tao/el-nino/nino-home-low.html

United States Geological Survey, http://www.usgs.gov/

United States Geological Survey, Climate Change and History, http://geology.usgs.gov/index.shtml

United States Geological Survey (USGS), Volcanoes in the Learning Web, http://www.usgs.gov/education/learnweb/volcano/index.html

CHAPTER 2

Earth's Biosphere

Earth's biosphere is one vast ecosystem. An ecosystem is a particular place, such as a forest or pond, where living and nonliving components interact to form a system.

ECOSYSTEMS

The living organisms or biotic parts of the ecosystem include animals, plants, fungi, protists, and bacteria. The nonliving components, or abiotic parts, include solar energy, water, soil, minerals, and gases such as oxygen and nitrogen. All organisms need energy and materials in their environment in order to live and survive.

Ecosystems vary greatly in size. A drop of pond water that contains living and nonliving parts may be considered to be an ecosystem.

The people in the park are part of an ecosystem. (Courtesy of Harry Cawthorn)

Neighborhoods, backyards, parks, lakes, and farms are all examples of other ecosystems.

BIOMES

A large ecological area characterized by similar vegetation and climate is called a biome. Climate is the main determining factor of a biome. A region's climate influences the kinds of plants that can grow which, in turn, determine the types of animals which have become adapted to live there. *Ecologists* name each biome according to the plant life that grows there.

Refer to Chapters 4, 5, and 6 for more information about specific biomes.

The six general types of land, or terrestrial, biomes include deserts, grasslands, taigas, temperate forests, rain forests, and tundras. The marine biome includes the ocean and seas. Freshwater lakes, wetlands, and rivers are aquatic biomes. There are no distinct boundaries between biomes. All of the biomes remain interconnected and interdependent.

LIVING ORGANISMS IN THE ECOSYSTEM

To understand how an ecosystem works, one must study the many various kinds of organisms. Any living thing is called an organism. Some of the main characteristics of living organisms include the ability to:

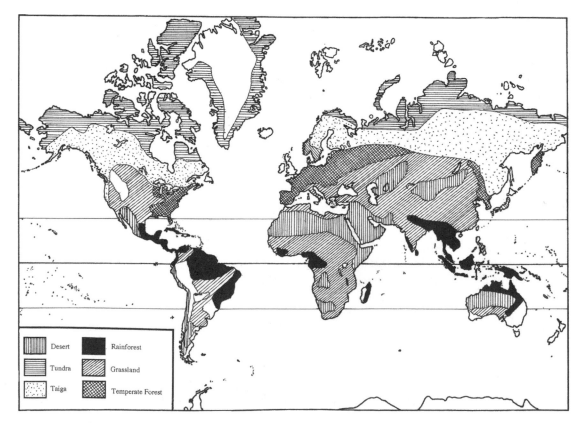

FIGURE 2-1 • The Major Land Biomes

- Adjust to changes in the environment

- Use energy

- Grow, develop, and reproduce.

A single kind of organism is called a species. Species are members of a group of organisms that can successfully produce offspring with each other under natural conditions. Each species includes all of the individuals that resemble one another and also differ in appearance from other groups. Therefore, even though the lion, house cat, and cheetah have similar catlike characteristics, these three cats belong to different species. The number of named species is about 1.6 million. However, it is estimated that between 5 and 30 million different species of organisms, which still have not been discovered, live on Earth. As an example, there are about 35,000 spider species. One scientist estimates that there may be as many as 135,000 more spider species that have not yet been discovered.

To better understand the many species, *taxonomists* organize and classify organisms into major kingdoms. Classification provides a framework to study the relationships among living and extinct species. One classification involves six kingdoms: archaebacteria, eubacteria, protists, fungi, plants, and animals. In this chapter, each of these groups will be described briefly.

Archaebacteria

There are approximately 500 species of archaebacteria, which are single-celled organisms that live in extreme and oxygen-free environments. They live in marshes, lake sediments, and in the digestive tracts of such mammals as cows. Archaebacteria can also live in a high concentration of saltwater, for example, in the Great Salt Lake in Utah, in the Dead Sea of the Middle East,

Native and Alien Species

Native species are species that have evolved, or originated, in a particular area. They have lived in certain locations for long periods of time and have become well adapted to their surroundings. Alien or nonnative species are species that are introduced to new locations. They can sometimes disrupt stable ecosystems by introducing new conditions such as disease to which the native species are not adapted. These conditions may make it impossible for the native species to coexist and live with newly introduced species.

Taxonomy

Taxonomy is the branch of biology that is concerned primarily with identifying and classifying the numerous organisms on the basis of their evolutionary relationships. Organisms studied by taxonomists may be living or extinct. To determine evolutionary relationships—how organisms are related to each other—taxonomists look at such features as cell structure and function, anatomy, similarities and differences among the structures of proteins and nucleic acids, and the sizes, shapes, and numbers of chromosomes. After all of these factors have been studied, organisms are placed into groups, called taxa, based on how their characteristics compare to those of other organisms.

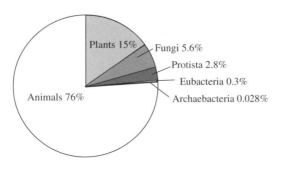

Plants 15%
Fungi 5.6%
Protista 2.8%
Eubacteria 0.3%
Archaebacteria 0.028%
Animals 76%

FIGURE 2-2 • The Percentage of Living Organisms in the Ecosystem

Viruses

Viruses generally are not considered to be organisms because they lack virtually all characteristics of living things. However, viruses depend on organisms for their existence because they are able to replicate (makes more of themselves) only when inside living cells. Viruses are of environmental importance because they often serve as agents of disease in plants, animals, and bacteria, and they are easily spread throughout dense populations of organisms. Some human diseases caused by viruses include Ebola virus, common colds, influenza (flu), measles, acquired immune deficiency syndrome (AIDS), herpes, tuberculosis, some forms of pneumonia, and even some forms of cancer. Thus far, scientists have identified more than 4,000 different viruses.

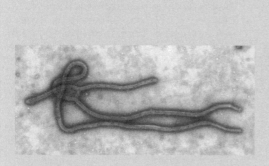

The Ebola Virus. (Courtesy of Centers for Disease Control and Prevention)

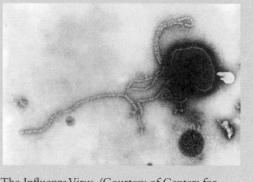

The Influenza Virus. (Courtesy of Centers for Disease Control and Prevention)

TABLE 2-1	Six-Kingdom Classification
Kingdom Archaebacteria	Bacteria that produce methane
Kingdom Eubacteria	Cyanobacteria, blue-green bacteria
Kingdom Protista	Algae with a nucleus, protozoa, and some fungi and slime molds
Kingdom Fungi	Usually form from spores, including some molds, mushrooms, and lichens
Kingdom Plantae	Plants
Kingdom Animalia	Animals

and in seawater-evaporating ponds. They can also live in hot acidic sulfur springs and in ocean vents in deep cracks of the Pacific ocean floor where temperatures can reach more than 100°C, the boiling point of water (212°F).

Eubacteria

There are more than 10,000 species of bacteria. Eubacteria, which are single-celled, microscopic organisms, are among the most abundant organisms on Earth. One type of eubacteria is called cyanobacteria, or

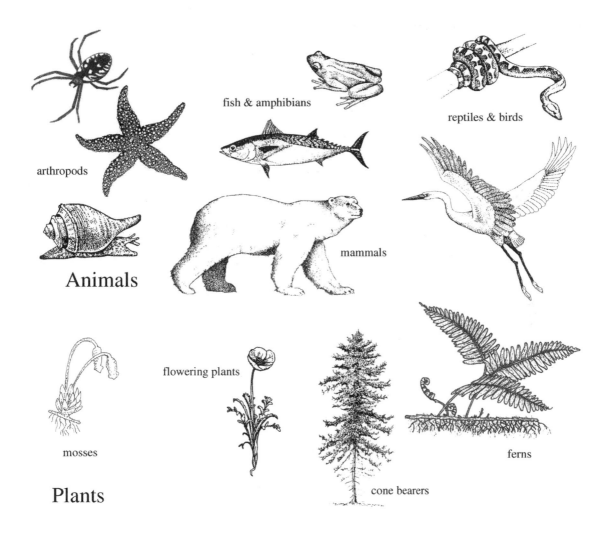

arthropods

fish & amphibians

reptiles & birds

mammals

Animals

mosses

flowering plants

cone bearers

ferns

Plants

mushrooms

molds

Protists

Fungi

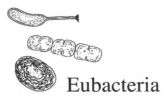

Archaebacteria

Eubacteria

Figure 2-3 • The Six Major Kingdoms

Archaebacteria live in high concentrations of salt water such as in the Utah's Great Salt Lake. (Courtesy of Salt Lake Convention & Visitors Bureau)

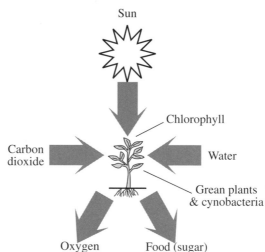

FIGURE 2-4 • **The Process of Photosynthesis**

Escherichia coli live in the intestinal tracts of humans. The intestinal tract provides Escherichia coli (E. coli) with a warm, safe environment and a constant food source. In return, the bacteria aid in the digestive process and help make vitamins needed by the human body. A similar relationship is seen with some bacteria that live in the intestines of cows, sheep, and horses. In these animals, the bacteria help the animals break down the cellulose contained in the plant matter eaten by the animals. However, E. coli outside of the human digestive tract causes disease in humans. Humans can get sick from E. coli by eating contaminated meat or drinking unsafe drinking water that has been polluted by plant or animal sewage.

blue-green bacteria. Like plants, these bacteria contain chlorophyll and are able to synthesize nutrients through photosynthesis. In this process, cyanobacteria take in carbon dioxide (CO_2) and water from the environment and return oxygen (O_2) as a waste product. Cyanobacteria can live in ponds, streams, moist places, and even rice paddies. Cyanobacteria provide food to many other organisms.

Another type of eubacteria manufactures nutrients via chemosynthesis. In this process, chemosynthetic bacteria lack chlorophyll and do not require sunlight for their food-making process. Instead, these bacteria make food by using chemicals such as carbon dioxide, sulfur, ammonia, iron, or nitrates. Chemosynthetic bacteria have been discovered

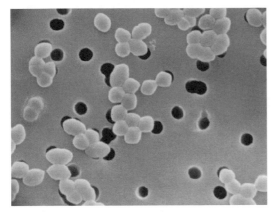

Round Bacteria. (Courtesy of Centers for Disease Control and Prevention)

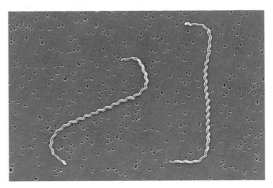

Spiral Bacteria. (Courtesy of Centers for Disease Control and Prevention)

TABLE 2.2	Diseases Caused by Bacteria

Disease	Causes	Symptoms
Strep throat	Inhale or ingest through mouth	Fever, sore throat swollen neck glands
Tuberculosis	Inhale	Fatigue, fever, night sweats, cough, weight loss, chest pain
Tetanus	Puncture wound	Stiff jaw, muscle spasms, paralysis
Lyme disease	Bite of infected tick	Rash at site of bite chills, body aches, joint swelling
Dental cavities (caries)	Bacteria in mouth	Destruction to tooth enamel, tooth ache
Cholera	Drinking contaminated water	Diarrhea, vomiting, dehydration

living in the deep ocean, near thermal vents in the ocean floor. The vents release chemicals, such as sulfate, which are used by these bacteria for chemosynthesis.

Many bacteria are important. They help fertilize crops and plants and recycle nutrients, and they are used to produce medicines and foods such as cheese.

Not all relationships between bacteria and other organisms are beneficial. Human diseases caused by bacteria include strep throat, Lyme disease, diphtheria, salmonella food poisoning, Legionnaire's disease, and tuberculosis. Bacteria also cause many diseases in plants and in animals other than humans. For example, anthrax and brucellosis are diseases of cattle and sheep, and sometimes humans, which are caused by bacteria.

Protists

There are approximately 60,000 species of protists. They are usually classified into different groups based on the ways in which they obtain nutrition and their mechanisms for locomotion. Some of the groups include protozoans, algae, and slime molds and downy mildews.

PROTOZOANS

The protozoans, or protozoa, are called animal-like protists because they are consumers and usually have structures enabling independent movement. Most protozoa live in moist habitats, including soil, freshwater lakes and ponds, and the ocean. One kind lives in the body of termites. The protists produce enzymes that digest wood, providing food for the termites. A few protozoan species cause disease in humans or other organisms. Examples include the amoebas and ciliated protozoans that cause the intestinal disorder known as dysentery. Dysentery, which can result from drinking water contaminated with protozoans, is most common in countries where drinking water is not treated.

ALGAE

Algae are plantlike protists; however, unlike plants, algae do not have organs such as roots, stems, and leaves. They are unicellular and multicellular protists that obtain their nutrients through photosynthesis and have cells enclosed within a cell wall. Common algae include the unicellular dinoflagellates that often cause red tides. Several species of dinoflagellates produce toxins that can be lethal. Multicellular protists include green algae which live in freshwater and red and brown algae which live in saltwater. The largest of the brown algae is kelp which can grow to more than 60 meters (180 feet). Kelp may form vast underwater forests that provide a habitat to a variety of marine species. Many brown algae have structures called air bladders which help them float near the water's surface, where they are exposed to the sunlight needed for photosynthesis. Brown algae generally live in cold ocean waters along rocky coasts.

As a group, algae are the main producers of aquatic ecosystems. They are also important for their role in releasing oxygen to the environment. It has been estimated that as much as 80 percent of the Earth's atmospheric oxygen is produced through photosynthesis carried out by algae.

SLIME MOLDS AND DOWNY MILDEWS

Another major group of protists include slime molds and downy mildews which are funguslike protists. Slime molds live in moist, shady places and come in a variety of colors ranging from bright yellow to black. Like fungi, many slime molds obtain their nutrients through decomposition, the breaking down of the remains of other organisms such as decaying logs and leaves. Thus, slime molds help to cleanse the environment of waste matter and also help to recycle nutrients through the environment. The downy mildews, which live in water or moist places, can cause disease in many plants. For example, the potato blight, which resulted in the Irish potato famine of the 1840s, was caused by a downy mildew that infected and destroyed the potato crop.

Fungi

Fungi belong in a kingdom of organisms of about 100,000 species which includes mushrooms, yeasts, rusts, smuts, molds, and mildews. As a group, fungi include both single-celled and multicellular species. Most fungi are decomposers, and they derive their nutrition by absorbing nutrients obtained through the decomposition, or breakdown, of organic matter, such as the wastes or remains of other organisms. In addition to their roles as decomposers and parasites, fungi are of environmental importance because they can be used in the treatment of disease. They also provide a source of food to many other types of organisms.

Fungi are widely dispersed throughout the world and exist in virtually every type of environment. In most ecosystems, fungi serve a vital role as decomposers—a role that benefits the environment in two ways. First, because they feed on organic waste materials, fungi help cleanse ecosystems of wastes that would otherwise build up on Earth's surface. In addition, as they feed, fungi break down complex organic matter into simpler organic and inorganic substances. Once broken down, matter that is not used by the fungi is returned to the environment, where it can be reused by plants and other organisms.

Most fungi thrive in habitats that are moist, dark, and warm. Such habitats include both soil and water. Many fungi are able to live in polluted habitats, where they derive nutrition by breaking down the organic matter present in the pollutants. For example, fungi have been found living in ponds and other aquatic habitats that are polluted with sewage.

Humans have found fungi useful in producing various products. For example, single-celled yeasts are used to make breads and alcoholic beverages. Other fungi have been used for many years by people around the world to make a variety of cheeses.

Fungi also produce a variety of substances that have medicinal value. Among the most well known is the antibiotic penicillin, which has been used worldwide to treat a variety of bacterial infections since the 1940s.

However, not all activities of fungi benefit other organisms. Many fungi are extremely toxic to humans and other organisms when ingested. Other fungi are pathogens, or agents of disease. Plant diseases

FIGURE 2-5 • The coconut scented milkcap is a fungus.

FIGURE 2-6 • Ferns are non-seed plants that reproduce by spores from its leaves and not from seeds.

caused by fungi include potato wart, rust and smut in corn, and club-root in cabbage.

Plants

The approximately 510,000 species of plants include nonseed plants and seed plants.

NONSEED PLANTS

Mosses, liverworts, and ferns are all nonseed plants. Mosses and liverworts are very small, many-celled plants which contain chlorophyll. They have no roots, stems, or leaves. They grow in mats of vegetation where there is plenty of water. Ferns, which are also nonseed plants, have true roots, stems, and leaves. Ferns reproduce by means of spores from their leaves and not from seeds. The spores grow into new plants. Ferns have no flowers.

FIGURE 2-7 • The Douglas Fir is a cone-bearing tree. It is not really a true Fir but belongs in the pine family. It is one of the largest and most valuable timber trees in the world. It is common in the western United States and Canada.

SEED PLANTS

Seed plants, which make up most of the plant populations in the world, consist of two major groups: gymnosperms and angiosperms. There are approximately 1,000 species of gymnosperms. Gymnosperms are the simplest seed plants and include evergreen trees or conifers. Conifers are cone-bearing trees that produce seeds in the cones. Angiosperms are the flowering plants, including more than 200,000 species of which the most common are wild and garden flowers and such hardwood trees as maple, birch, and oak. Other seed-producing plants include cycads, ginkgoes, climbing vines, and some shrub like species.

Animals

The animal kingdom includes about 2,900,000 species. The grouping of animals in the animal kingdom include invertebrates and verte-brates. Invertebrates are animals that do no have an internal backbone, such as shellfish and insects. Vertebrates are animals that have an

Figure 2-8 • Some invertebrates include mollusks, barnacles, snails, and spiders

internal backbone composed of vertebrae. Vertebrates include such animals as birds, reptiles, and mammals.

INVERTEBRATES

Invertebrates include such animals as sponges, flatworms, mollusks, and arthropods. Of all the invertebrates, arthropods are probably the most familiar. About 80 percent of all animal species are arthropods. They include spiders, insects, centipedes, millipedes, and scorpions.

Crustaceans are a class of arthropods containing about 35,000 species. Crustaceans are an important component of food chains and food webs. As arthropods, crustaceans are invertebrates with jointed appendages and an exoskeleton composed largely of chitin. Members of this class, including shrimps, crabs, water fleas, and woodlice, are distributed throughout the world, primarily in freshwater and marine habitats. Woodlice are an example of terrestrial crustaceans. Many crustaceans, especially shrimps, lobsters, and crabs, are an important food source for people. Others, such as the krill that are a component of plankton, serve as a major food source for marine organisms, including many species of whales.

VERTEBRATES

REPTILES AND BIRDS Reptiles are a class of vertebrates comprising approximately 5,000 species including snakes, lizards, alligators, crocodiles, turtles, tortoises, and the rare tuatara. Reptiles are characterized as being *ectothermic*, primarily egg laying, with a body covering of scales or horny plates. Most reptiles live on land and breathe using lungs, although a few species of snakes and turtles are adapted to life in freshwater or marine environments. As a group, reptiles are widely distributed throughout the world and live in virtually all environments, except those having arctic or subarctic climates.

Birds are the only animals that have feathers and, unlike reptiles, a bird is a warm-blooded animal. Birds can be divided into two groups: those that fly and those that do not. Flightless birds include ostriches and rheas.

DID YOU KNOW?

Within the insect species, the order *Coleoptera* (beetles) is the largest. There are more species of beetles than of any other species on Earth.

MAMMALS A large class of *endothermic* vertebrates that have a constant body temperature and whose offspring are nourished by milk from the mother are mammals. Other characteristics of mammals include body hair, live birth in all but one group of mammals, and outer ears. Mammals also differ from other animals in that they have a flat muscle called a diaphragm which separates the chest from the abdominal cavity.

Mammals, which total about 4,500 species, are widely distributed today in varied environments including deserts, forests, grasslands, and tundras as well as in the waters of the ocean. Mammals are classified into three main groups: egg-laying mammals, or monotremes, including the platypus and spiny echidnas; marsupials which have pouches in which the young develop, including the kangaroo and opposum; and placental mammals, in which the young develop inside the body of the female and are born alive. The placental mammals are further classified into orders that include flying mammals, such as bats; hoofed mammals,

FIGURE 2-9 • The zebra population throughout the world is decreasing because of loss of habitats, hunting, and political instability.

FIGURE 2-10 • The sea lion is protected under the Marine Mammal Protection Act of the United States.

such as rhinoceroses, horses, and deer; and mammals that live in the ocean, such as blue whales and manatees.

AMPHIBIANS AND FISH Amphibians, or cold-blooded, egg-laying vertebrates such as frogs, toads, and salamanders, are declining in population in parts of Australia, the western United States and Canada, Central America, and South America. Amphibians are a class of vertebrates whose characteristics include a three-chambered heart, cold bloodedness, and a lack of hair, scales, or feathers. Many species lay their eggs in water. After hatching, the young are herbivores which feed on plants and draw oxygen from the water through gills. Later, the amphibians develop lungs, become *insectivores*, and live on land as adults. About 4,000 species of amphibians are found in many ecosystems, including deserts, forests, and grasslands, and at various altitudes. They are abundant in the tropics but can also be found in the temperate zones and in the higher latitudes of northern North America.

Unlike amphibians, fish spend all their lifetime in freshwater or saltwater habitats. Fishes form three classes of vertebrates of which there are 20,000 species. They include jawless fish such as lampreys, cartilaginous fish such as sharks, and bony fish such as swordfish.

Diversity of Living Organisms

The variety and abundance of differing wildlife organisms living in an area of the biosphere is known as species diversity. The term also means the richness of the number of species. As an example, a rich diversity of organisms would include those species that live in a rain forest, grasslands, or coral reef. In these environments, one would find a large abundance and varied group of organisms. On the other hand, in a desert the same size of a rain forest, one would find a less abundant, less varied group of organisms.

ADAPTATION OF SPECIES

Any trait of an organism that makes it suited to life in its environment is called an adaptation. Adaptations include features of an organism such as its body covering, its color, and its physical structures for movement and

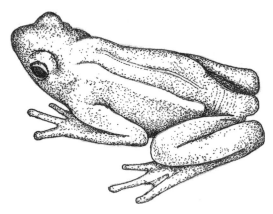

FIGURE 2-11 • The golden frog is an amphibian. Amphibians are able to live both on land and in the water.

FIGURE 2-12 • The wild salmon have been threatened by a variety of conditions. Excessive nitrogen from animal waste and fertilizers has reduced the dissolved oxygen in streams and rivers.

Keystone Species

Keystone species are species whose presence or absence in an ecosystem significantly impacts the other species in that ecosystem. Scientists have determined that removing a keystone species from an ecosystem may significantly reduce the biodiversity of that ecosystem. The sea otter that inhabits the kelp forests of the Pacific Ocean off the coasts of the United States and Canada has been identified as a keystone species. Submarine kelp forests are areas of dense kelp growth that provide habitat or breeding grounds to numerous fish and wildlife species. During the early twentieth century, sea otters were greatly overhunted, decreasing their population size. In a short time, scientists also observed a significant decline in the kelp that constituted the ecosystem in which these animals had thrived. Studies of the region indicated that decimation of the kelp resulted from a population explosion of the sea urchins on which the sea otters fed. Without a predator to keep their numbers in check, the sea urchin population thrived and exploded, while the kelp and seaweed populations on which the sea urchins fed greatly declined. The thinning of the kelp population not only removed a major food source, but also destroyed the habitat of many organisms, forcing them to move out of the area. This once diverse undersea ecosystem was transformed into a territory that was barren of life. To reverse the problem, the sea otter was reintroduced to the region and provided protected status under the Marine Mammal Protection Act of 1972. Since their reintroduction, the kelp forests have recovered and once again serve a thriving ecosystem.

The praying mantis has structural adaptations which it uses to get its food. As an example, it has mouthparts adapted for holding, chewing, and biting various foods. (Courtesy of David B. G. Oliveira, PhD., FRCP)

food getting. The praying mantis has structural adaptations which it uses to get its food. Other adaptive features may include nest building, migration, courting rituals, whether the organism is able to carry out photosynthesis, or whether the organism is ectothermic or endothermic.

Each species has traits that make it adapt to life in a particular environment. Such adaptations result from evolution as organisms compete with one another in response to changing environmental conditions. In this process, nature selects those individuals that possess the traits that make them the most suited for survival in their environment. These

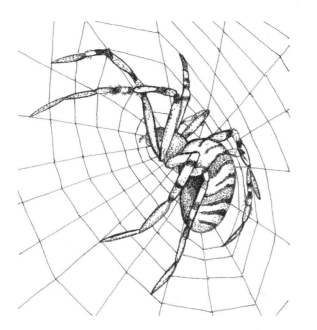

FIGURE 2-13 • Many spiders are predators who use their webs to catch prey. However, not all spiders spin webs.

organisms, in turn, are most likely to reproduce and thus pass these beneficial traits on to their offspring. Scientists use the term natural selection to describe the process by which certain traits in species appear or disappear as the environment favors individuals that produce the greatest numbers of surviving offspring. This process was first described by naturalist Charles Darwin in 1859.

Living things interact with other organisms. The interaction includes such relationships as predation, competition, mutualism, commensalism, and parasitism.

Predation

An ecological interaction between organisms of different species in which one organism hunts, kills, and eats the other as a means of obtaining nutrition is know as predation. In this relationship, the organism that does the capturing and eating is the predator, while the organism that is eaten is called the prey. Because they feed on other animals, predators are carnivores. Examples of predation exist in all ecosystems. For example, a spider that uses a web to capture insects is a predator, while the captured insects are its prey. The spider, in turn, may become the prey of another animal, such as a bird, which captures it for use as food. Predation helps to maintain balance within ecosystems. Predators are most likely to capture and feed on old, weak, or sick members of a population.

Competition

An interactive relationship present in all ecosystems, which occurs when two or more organisms, populations, or species try to use the same resources is known as competition. Examples of resources for which animals compete include food, water, living space, social status, and mates. Plants compete for water, soil nutrients, and sunlight. Competition can occur when wolves and coyotes try to use the remains of the same animal as food. It also occurs when small plants growing near ground level compete with taller plants for sunlight, water, and minerals in the soil.

Mutualism

A close association between two organisms of different species that benefits one or both organisms is called symbiosis. Mutualism is a symbiotic association between two organisms of different species through which both species derive some benefit. An example of such a relationship is seen in lichens. Lichens are an association between a fungus and either a green alga or a cyanobacterium. In this relationship, the fungus provides the alga or cyanobacterium with shelter and raw materials, such as carbon dioxide and water, needed for photosynthesis. The alga or cyanobacterium provides the fungus with a source of nutrients. In this case, the organisms are so interdependent that they are considered to be a single organism.

Lichens can live almost anywhere—from the coldest to the hottest places on Earth. Their ability to grow in many places makes them very good indicators of pollution. Lichens absorb minerals, nutrients, water, and other substances from both solid and liquid substances and from the air around them. However, lichens cannot excrete the unnecessary substances, some of which are toxins, from the air, water, and soil. The absorption of these substances in high amounts can cause the deterioration and the breakdown of the photosynthesizing unit of the lichens. One such toxin is sulfur dioxide (SO_2), a major component of polluted air. Many lichens are too sensitive to exist in areas where sulfur dioxide is present in the air, or if they exist, they do so in small populations. Therefore, lichens can be used as indicator species to monitor and gauge air-pollution levels in many areas.

Besides lichens, many plants have a mutualistic relationship with fungi. Mycorrhizae are mutualistic associations occurring between plant roots and fungi. The fungi benefit by absorbing nutrients made by the plant through photosynthesis; the plants benefit as the fungi decompose organic matter in soil, which the plants absorb as nutrients. In addition, the hyphae (threadlike structures) of the fungi increase the surface area of the plant's roots, enhancing the plant's ability to absorb water and nutrients in soil. Plants having mycorrhizae on their roots generally grow larger than those lacking the association.

Many mutualistic symbiotic relationships are obligatory because the organisms involved cannot survive without each other. For example, plants called legumes have nodules on their roots in which *Rhizobium* bacteria live. The bacteria carry out nitrogen fixation, in which nitrogen gas from the air is changed into a form of nitrogen that

Lichens are a good example of a mutualism association. Lichens are an association between two organisms of different species—a fungus and either a green alga or cyanobacterium. (Courtesy of Harry Cawthorn)

the plant can use as a nutrient. The bacteria benefit from the habitat provided by the nodules. The plant benefits from the nutrients provided by the activities of the bacteria.

Commensalism

Commensalism is a type of symbiotic association between two species of organisms in which one species derives benefit, while the other neither benefits nor is harmed. Many commensal relationships exist in nature. Often these relationships provide the benefiting species with increased access to resources such as food, water, and sunlight. An example of a commensal relationship which provides a species with greater access to food exists between barnacles and some species of whales. In this relationship, the barnacle, which is a sessile, filter-feeding animal, attaches to a whale. As the whale swims, water moves over the barnacle, carrying with it the small microorganisms the barnacle uses as food. The relationship between barnacles and whales also aids in the colonization of a greater portion of the ocean by barnacles. Many species of orchid are epiphytes which grow in the canopy of tropical rain forest trees. The trees provide the orchids with a habitat which allows them to obtain adequate sunlight, while also being exposed to the air from which they obtain the water and minerals they need for photosynthesis and growth. The trees do not appear to be affected by the relationship. To summarize, in a commensal symbiotic relationship, only one species benefits from the association. The other species neither benefits nor is harmed.

FIGURE 2-14 • The relationship between a whale and an attached barnacle is an example of commensalism.

Parasitism

Parasitism is a relationship between two species of organisms in which one, the *parasite*, lives in or on the other, called the host, as a means of obtaining nutrition. Every group of organisms—bacteria, protists, fungi, plants, and animals—has some parasitic species, and every species of organism can serve as host to some type of parasite. Thus, parasitism can be present in any ecosystem.

The effects of parasitism on a host may range from insignificant or minor to severe illness or, in rare cases, death. Typically, as a parasite derives nutrition, its actions weaken the host organism, without causing its death. Ectoparasites derive their nutrition by feeding on hair, feathers, scales, skin, or the blood of their host. Examples of such parasites include fleas, ticks, some flies, lice, and mosquitoes. Generally, these parasites must be present in large numbers to harm their host significantly by robbing it of nutrients; however, a single ectoparasite may harm its host indirectly by transmitting disease. Lyme disease, malaria, African sleeping sickness, and Rocky Mountain spotted fever are examples of diseases transmitted to humans by ectoparasites.

DID YOU KNOW?

Ticks obtain nutrients from the animals they live on. This relationship is called parasitism. Ticks can cause harm to the host animal. As an example a deer tick carries Lyme disease.

POPULATIONS AND COMMUNITIES

Populations

A population consists of a group of organisms of the same species that live, grow, and reproduce in the same environment. For example, all of the dandelions on a lawn make up a population. The population of bass in a lake includes all of the bass.

Population studies are extremely important to ecologists. They provide an indication of the overall health and stability of ecosystems. As a rule, ecosystems comprising many diverse populations are found to be more environmentally stable than those comprising only a few species. For example, if an ecosystem comprises only four or five species (populations) of organisms, it is likely that the removal of any one species would significantly affect all the others, since its removal would likely eliminate the main food source for at least one of the

FIGURE 2-15 • An elephant herd is a *population* consisting of organisms of the same species.

other populations. This population would quickly decrease in size as its members became unable to meet their food needs.

POPULATION DENSITY

The number of individuals of a given population that are living in an particular place at the same time is known as the population density. For example, if a total of 300 dandelions live on a square meter of the lawn, the population density of dandelions in the area would be indicated as 300 dandelions/square meter. If another area of the same size had a population of 600 dandelions, its population density would be 600 dandelions/square meter—a population size twice that of the first area. Studies of population density provide environmental scientists with important data about the overall health of ecosystems.

Every ecosystem has a carrying capacity, or a maximum number of individuals, for each population that it can support. The carrying capacity is dependent upon the availability of food, light, water, heat, and shelter. If these conditions are favorable, the population remains stable or increases. However, when the conditions are not favorable, the population will get smaller. Studies of population density provide scientists with an indication of the carrying capacity for each species of a given ecosystem. Population density studies also can alert scientists as to when some species might be at an increased risk of predation, parasitism, or disease.

Communities

Animals and plants and other living organisms seldom live alone. Many different species live together in a given area or community. A community is a collection of several interacting populations which live in a common environment. For example, the organisms of a pond may be referred to as a pond community. Such organisms include all the animals and plants living within or at the surface of the pond, including

FIGURE 2-16 • The elephant herd lives together with other species in a given area of *community*.

any microorganisms such as bacteria, protists, and fungi that are present. Communities may also be named for their dominant species, as in a pine forest community, in which the dominant species are pine trees. Other organisms of this community would likely include animals such as deer, squirrels, toads, salamanders, earthworms, and rabbits; plants such as mosses and ferns; fungi such as bracket fungi, puffballs, molds, and mushrooms; and bacteria and protists that live in the soil, air, water, and on or in other forest organisms.

HABITAT

The habitat is an environment, or place, where an organism normally lives. It provides the organism with all the biotic factors and abiotic factors—soil, food, water, shelter, proper temperature, light, moisture, mates—it needs to sustain life, as well as to ensure survival of its species.

A habitat can be thought of as an organism's home. Thus, the habitat of an individual organism may be described in very specific terms, such as a particular tree in a particular forest. However, when discussing a species, habitat is often identified in broader terms which describe an entire ecosystem or biome. These broader terms are generally descriptive of the dominant plants or physical features of an area. For example, a hardwood forest is a habitat in which the dominant plants are broadleaf, deciduous trees such as oak, maple, and birch. Such a forest provides habitat not only to the trees, but also to a variety of other plant species as well as animals, fungi, protists, and bacteria. The presence of water, as in a bog, lake or pond, hot spring, or ocean, is a physical feature that may be used to identify a habitat. Such environments provide habitat not only to the organisms living in the water, but also to the terrestrial organisms living nearby that interact with the water or its organisms.

NICHES

The term niche is used by ecologists to describe the role of an organism in its ecosystem. A niche also may describe the physical location or the lifestyle of the organism in its environment. For example, two species of *Anolis* lizards, which live in the same parts of the tropics, feed exclusively on insects. However, each species feeds on insects of different sizes. *Anolis* lizards with small jaws feed only on small insects, while similar lizards with large jaws feed exclusively on large insects. Because they have slightly different food sources, the niches of the two types of lizards are different, even though they share the same habitat.

In the case of the *Anolis* lizards, the large-jawed lizard could feed on insects of all sizes. Thus, its fundamental niche, the niche it is able to occupy, comprises both large and small insects. However, the fact that they do not feed on small insects reduces the competition for food between the small-jawed *Anolis* and the large-jawed *Anolis*. Thus the large-jawed lizard feeds almost exclusively on large insects. Feeding on large insects is

FIGURE 2-17 • The saguaro is part of the woodpecker's niche.

Zoology

The branch of biology that deals with the study of animals and animal life, including their roles in the environment, is known as zoology. Scientists who work in this field are called zoologists. One of the major concerns of zoology is animal classification, or taxonomy. Animals are classified into taxonomic groups, or taxa, based on their traits and evolutionary histories. Other major concerns of zoology are anatomy, physiology, and animal development.

the niche of the large-jawed *Anolis*. The tendency for two similar species to divide a niche between them is sometimes called niche splitting or niche differentiation. This tendency reduces competition and helps each species meet its needs. Some animals, such as rats live in broad niches while others, such as the panda, live in narrow niches. The rat can feed on a variety of different foods provided in many niches. The panda depends on one kind of food and is limited to one niche.

Organisms of the same species generally have the same niche in an ecosystem; however, organisms of different species cannot occupy the exact same niche at the same time. In other words, two populations cannot occupy the same niche. If this happens, one species generally will outcompete the other, forcing the less-adapted species to occupy a slightly different niche within the same ecosystem, move to another ecosystem where its niche is not already occupied, or risk extinction.

Vocabulary

Ecologists Scientists who are very concerned with how the biotic (living organisms) parts of the environment interact with each other as well as with the abiotic (nonliving) parts.

Ectothermic A term used to describe organisms whose metabolic processes are not sufficient to regulate body temperature and thus have body temperatures that are largely controlled by the surrounding environment. Such organisms, often termed ectotherms, include animals classified as fishes, reptiles, and amphibians.

Endothermic A term used to describe organisms whose body temperatures are regulated by

their metabolic processes. Such organisms, often termed endotherms, include animals classified as birds and mammals.

Insectivores Animals that eat insects.

Parasite Term derived from the Greek word *parasitos*, which means "one who eats at the table of another." The organism benefits at the expense of the other species.

Taxonomists Biologists who study taxonomy. Taxonomists group and name organisms based on studies of their shared characteristics.

Activities for Students

1. Create a poster that charts the six kingdoms of living organisms. Be sure to add illustrated examples for each of the kingdoms.

2. Choose an animal that lives in extreme conditions. Consider what adaptations it has made to survive. How have these adaptations helped the animal establish a niche?

3. Spend some time in your backyard or a local park. Observe the different populations that exist in that habitat. What types of relationships are there in the interactions between the various organisms?

Books and Other Reading Materials

Aber, John D., and Jerry Melillo. *Terrestrial Ecosystems.* New York: Harcourt/Academic Press, 2001.

Art, Henry W., ed. *The Dictionary of Ecology and Environmental Science.* New York: Henry Holt, 1993.

Cain, A. J. *Animal Species and Their Evolution.* Princeton, N. J.: Princeton University Press, 1993.

Cogger, Harold G., and Richard G. Zweifel, eds. *Encyclopedia of Reptiles & Amphibians.* San Diego, Calif.: Academic Press, 1998.

Hare, Tony. *Animal Habitats: Discovering How Animals Live in the Wild.* New York: Facts on File, 2001.

Holley, Dennis. *Viruses & Bacteria: Hands-On Minds-On Activities for Middle School & High School.* Critical Thinking Books and Software, 1999.

Hunkin, Jorie. *Ecology for All Ages.* Guilford, Conn.: Globe Pequot Press, 1994.

Margulis, Lynn, and Karlene V. Schwartz. *Five Kingdoms: An Illustrated Guide to the Phyla of Life on Earth.* 2d Ed. New York: W. H. Freeman, 1996.

Martin, Glen. *National Geographic's Guide to Wildlife Watching: 100 of the Best Places in America to See Animals in Their Natural Habitats.* Washington, D.C.: National Geographic Society, 1998.

Tompkins, Peter. *The Secret Life of Plants.* New York: HarperCollins, 1989.

Watkinson, Sarah, Michael Carlile, and Gooday Graham. *The Fungi.* 2d Ed. San Diego: Academic Press, 2001.

Websites

Green Teacher, ideas for environmental projects, http://www.greenteacher.com/

National Wildlife Federation, observing backyard habitats, http://www.nwf.org/habitats/schoolyard/

Natural History and Ecology for Homo Sapiens, suggestions for projects, http://www.accessexcellence.org/

NOAA Education Website, Specially for Students, http://www.education.noaa.gov/

USDA for Kids, U.S. Department of Agriculture, http://www.usda.gov/news/usdakids/index

U.S. Fish and Wildlife Service, The Species List of Endangered and Threatened Wildlife, http://www.fws.gov/r9endspp/lsppinfo.html

U.S. Fish and Wildlife Service, Vertebrate Animals, http://www.fws.gov/r9endspp/lsppinfo.html

Welcome to One Sky Many Voices, Environmental science programs and projects, http://www.onesky.umich.edu/

CHAPTER 3

How Ecosystems Work

Every organism, including humans, takes in *energy* to survive. Some of the energy is used for growth, reproduction, and body repair. Organisms also need and take in *matter* in the form of chemical elements and nutrients found in water, soil, and air. Ecosystems work when there is a continuous supply of energy and matter.

PRODUCERS

The living organisms of an ecosystem include producers, consumers, and decomposers. A producer, also called an autotroph, is an organism that is capable of making its own food. Producers are a vital component of all ecosystems because all other organisms derive their energy (via food) either directly or indirectly from producers.

Photosynthesis

Producers that manufacture their food via photosynthesis include all plants, algae, and some other protists, as well as the bacteria known as cyanobacteria. Green plants are producers that use *pigment*, such as chlorophyll, the green pigment in leaves, to capture the energy in sunlight. The sunlight drives a chemical reaction which combines carbon dioxide (CO_2) and water (H_2O) to make glucose ($C_6H_{12}O_6$), a sugar that is used as food and energy by the producer. The oxygen (O_2) formed by this reaction is released into the environment as a byproduct.

Glucose, the end product of photosynthesis, is used for making proteins and carbohydrates. A portion of the glucose is stored in the plant's cells or tissues as starch. A starch is a *carbohydrate* composed of simple sugars. It is a major component of many different kinds of foods including rice, potatoes, wheat, and cereal grains. Starch represents 70 percent of the world's food supply.

Animals that get their energy and food by feeding on producers fall into a group ecologists term "primary consumers." As an example, a mouse that is eating plant material is getting its energy from the glucose in the food that the plant made during photosynthesis. The mouse may use some of the glucose contained in the starch for its life processes or store unused glucose in its cells, tissues, and organs.

DID YOU KNOW?

Photosynthesis occurs only during the daylight hours. Special indoor lights can also be used to encourage photosynthesis.

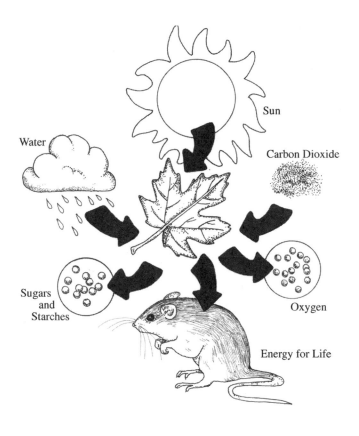

FIGURE 3-1 •
Photosynthesis In this diagram, photosynthesis occurs when a plant uses the energy in sunlight to drive a chemical reaction that converts carbon dioxide (CO_2) and water (H_2O) into simple sugars and starches. The plant provides most other organisms, such as the mouse, with food and energy. Photosynthesis also releases oxygen into the environment as a byproduct.

Phytoplankton

Green plants such as trees and shrubs are familiar land producers; however, some aquatic organisms are also very important producers. Phytoplankton is one of the major producers of oxygen and nutrients in an aquatic ecosystem. These tiny aquatic unicellular organisms, mostly algae, derive their nutrition and energy through photosynthesis.

Phytoplankton generally lack structures that enable independent movement, and instead they drift at or near the surface of the water, where they are exposed to sunlight. Movement of these organisms, which include photosynthetic bacteria and algae, generally results from movement of the water (winds, tides, or currents) in which they live. As producers, phytoplankton are the primary food source for most zooplankton. Phytoplankton also are the main food for many of Earth's largest aquatic organisms, including several species of whales. Other aquatic organisms that are producers include multicellular species of red, brown, and green algae.

Chemosynthesis

Not all producers carry out photosynthesis to obtain their food. Some use a process known as chemosynthesis. Such organisms get the energy needed to drive their food-making process by oxidizing or reducing chemicals, including inorganic and organic compounds such as hydrogen sulfide (H_2S), methane (CH_4), and ammonia (NH_3). Organisms

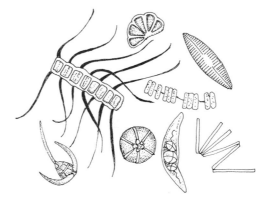

FIGURE 3-2 •
Phytoplankton are of great environmental importance because they form the base of the food chains and food webs.

that obtain their nutrition in this way are known as chemotrophs or chemoautotrophs. These organisms are made up of a small number of bacteria and a few species of protozoans. Some are aerobic. Some are anaerobic. Chemosynthetic organisms have been found living near deep ocean vents in areas that cannot be reached by sunlight. These organisms provide food for a variety of unusual organisms, such as tube worms, some of which are as long as 1 meter (3.3 feet).

CONSUMERS

A consumer is an organism that cannot synthesize its nutrients from energy or chemicals and must therefore obtain energy by consuming other organisms. Consumers are also known as heterotrophic. Consumers include all animals and fungi and those species of protists and bacteria that do not carry out photosynthesis or chemosynthesis. Primary consumers feed on producers, other consumers (called secondary consumers), or both. Most consumers are classified as herbivores, carnivores, or omnivores, on the basis of the types of organisms they eat.

Herbivores

Herbivores, or plant eaters, which are primary consumers, feed only on producers and their products. Examples include grazing animals, such as horses, cattle, bison, and the zebras of the African plains, as well as browsers, such as giraffes. Many kinds of insects, seed-eating birds, algae-eating fishes and shellfishes, and fruit-eating monkeys and bats are also herbivores.

Carnivores

Carnivores, or flesh eaters, which are secondary consumers, feed on other consumers. Many carnivores are meat eaters; however, carnivores may feed on such nonmeat products as eggs and milk as well. Meat-eating

FIGURE 3-3 • The wildebeest, a herbivore, lives in Tsavo National Park in Kenya, which is a refuge for many endangered elephants and other threatened species.

FIGURE 3-4 • The cheetah is a carnivore. Namibia, Africa, has the largest population of cheetahs.

carnivores are predators that hunt and kill *prey* for food. Some of the top carnivores are lions, polar bears, sharks, falcons, and toothed whales. Spiders are the major predators of the insect population. Studies concur that a spider can eat more than 100 insects in a year.

A few plants also feed partly on meat. These include Venus's-flytraps, butterworths, sundews, and pitcher plants. These plants trap their prey, mainly insects, by producing a sticky substance that prevents the prey from escaping. The plant ingests the body fluids of the prey.

Omnivores

Omnivores, or eaters of all, are consumers that feed on both plant and animal products. Human beings, bears, coyotes, pigs, and raccoons, as well as many kinds of birds and reptiles, are omnivores.

Scavengers and Decomposers

Scavengers and decomposers are other types of consumers. These organisms feed on detritus or the wastes or remains of other organisms, such as dead trees, dead grass, fecal wastes, and dead animals. Examples of scavengers include vultures and hyenas. Because they feed on meat products, these scavengers are carnivores; however, they differ from other carnivores in that they do not kill the organisms they eat. Instead, they feed on the remains of animals that have died in the wild, or carrion, or on animal parts left behind by predators.

The majority of decomposers are bacteria and fungi. Decomposers obtain their nutrients by breaking down the wastes or remains

FIGURE 3-5 • Fungi are commonly observed growing in moist, shady areas.

of other organisms into simpler substances they can use as food. Although scavengers and decomposers use similar sources for their food, the two types of organisms differ in how they obtain their nutrients. Decomposers release chemicals that break down or digest matter before it is absorbed or taken into the "body." Scavengers ingest the matter before it is broken down through digestion.

Both decomposers and scavengers play an important ecological role. They clear the wastes and remains of other organisms from the environment. Decomposers also help to recycle nutrients back into the environment, especially the soil, where the nutrients can be used by plants and other organisms.

ENERGY FLOWS THROUGH ECOSYSTEMS

Food Chain

A model used by ecologists to trace the flow of energy and nutrients through an ecosystem, based on the feeding patterns of its organisms, is known as a food chain. A food chain traces only one pathway for the transfer of energy and matter through an ecosystem. Thus a single ecosystem generally has many different food chains. All food chains have some elements in common. For example, all food chains begin with producers. Additional organisms in a food chain include one or more levels of consumers and conclude with decomposers.

Because the sun is the major source of energy for most ecosystems, the producer at the base of a food chain is usually an organism that derives its nutrients through photosynthesis. In terrestrial ecosystems, the producers are usually plants, whereas photosynthetic producers in aquatic ecosystems are often phytoplankton and algae. In some ecosystems, such as ocean vent communities of the deep ocean, the producers at the base of the food chain are bacteria that obtain their nutrients by synthesizing inorganic materials in the process of chemosynthesis.

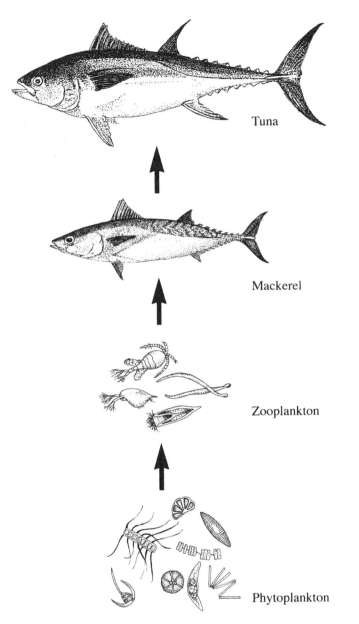

Tuna

Mackerel

Zooplankton

Phytoplankton

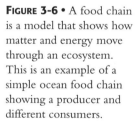

FIGURE 3-6 • A food chain is a model that shows how matter and energy move through an ecosystem. This is an example of a simple ocean food chain showing a producer and different consumers.

Feeding Levels

Each feeding level in a food chain is known as a *trophic level*. The bottom or first trophic level in all food chains is made up of producers such as green plants or algae. A consumer, known as the primary consumer, or first-order consumer, makes up the second trophic level of a food chain. The primary consumer, such as a herbivore, feeds directly on a producer. As it feeds, the primary consumer obtains both nutrients and energy stored in the producer. The primary consumer may then be eaten by a second-order consumer (a carnivore or omnivore), which makes up the third trophic level of the food chain. As in the second trophic level, the secondary consumer obtains its nutrients and energy from the primary consumer it eats. The secondary consumer may be prey for a top predator, or third-order consumer, such as an eagle or polar bear. The top predator is a tertiary consumer that makes up the fourth trophic level. This predator (a carnivore or omnivore) generally does not serve as prey to any other consumers in the ecosystem. However, when this predator does die, it may become a source of food for a scavenger or be consumed directly by decomposers (bacteria and fungi). A food chain may consist of three trophic levels, but there are no more than five links before ending with a top predator.

Food chains are often represented in visual models in which arrows show the movement of nutrients and energy from one organism to the next. A possible food chain for a meadow comprises grass (producer), mouse (primary consumer), snake (secondary consumer), and eagle or bacteria (decomposers).

Food Webs

A food web is a model comprising several overlapping food chains in an ecosystem. It has a shape like a web. Ecologists can trace the flow of energy and nutrients through an ecosystem by drawing lines on the model between organisms that feed on each other. All ecosystems have some organisms that are classified as consumers—organisms that obtain their food and energy by eating other organisms. Often, consumers use more than one kind of organism as a food source. Thus, many consumers are involved in more than one food chain. A food web provides ecologists with a mechanism for tracing multiple pathways involving the transfer of energy and nutrients at the same time.

A food web includes all of the same trophic levels, or feeding levels, as a food chain. However, unlike a food chain, a food web illustrates how a single organism may feed at more than one trophic level. For example, a field mouse is an omnivore that sometimes feeds on seeds and sometimes feeds on insects. When feeding on seeds, the mouse is feeding directly on a producer, and thus is a primary consumer eating at the first trophic level. However, if this same mouse eats a grasshopper (which feeds on plants), it is a secondary consumer feeding at the second trophic level. A food web can show both of these feeding relationships at the same time.

Food webs can be represented in visual models that use arrows to show the movement of energy and nutrients from one organism to the next. For example, in one food web, for a deciduous forest ecosystem, an oak tree (producer) is the food source of a squirrel (primary consumer), which may then be eaten by a fox (secondary consumer), which is a top predator in this ecosystem. In addition, a squirrel (primary consumer) may be eaten by a snake (secondary consumer), which is then eaten by a tawny owl (tertiary consumer), which is also a top predator in this ecosystem. These overlapping food chains

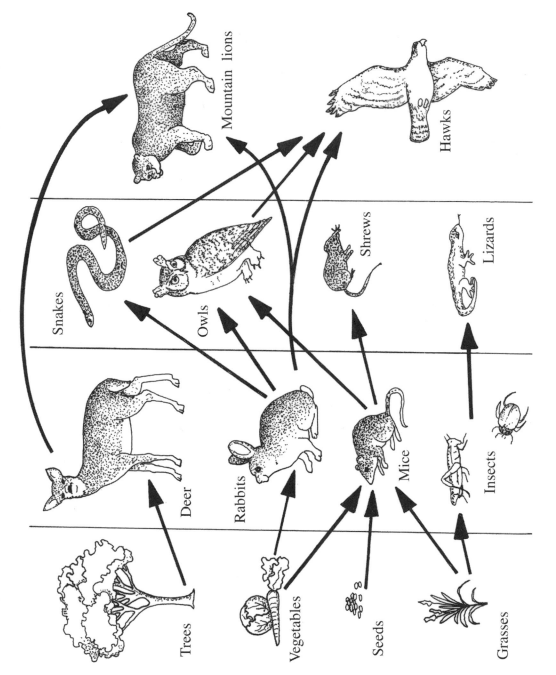

Producers Primary consumers Secondary consumers Tertiary consumers

Mountain lions

Hawks

Snakes

Owls

Shrews

Lizards

Deer

Rabbits

Mice

Insects

Trees

Vegetables

Seeds

Grasses

FIGURE 3-7 • Food Web A food web is a model that expresses all the possible feeding relationships at each trophic level in a community. This food web shows several feeding patterns involving producers and consumers of typical neighboring grassland and deciduous ecosystems.

involving the oak tree and the squirrel provide scientists with more information about the interactions among the organisms in this ecosystem than does a single food chain. Such a food web also shows several other feeding relationships which may involve these same organisms.

Pyramid of Energy

To display how energy moves within an ecosystem, scientists use a model known as a pyramid of energy. The model shows the relative amounts of energy present at the different trophic levels within an ecosystem. The base of an ecological pyramid represents the first trophic level, or producers, of an ecosystem. This trophic level contains the greatest amount of energy available to the ecosystem and is assigned a value of 100 percent. Above the base is a level representing the primary consumers. These organisms feed on producers or their products to obtain their food and energy. Because some of the energy present at the producer level is used by the producers to carry out their life processes, less energy is available to the organisms constituting the second trophic level. Thus this level is shown as being smaller in size than the producer level. At each higher level there is less energy available. In fact, scientists estimate that only about 10 percent of the total energy available at one level is passed to the next higher trophic level through the food chain. The remaining 90 percent is either used by the organisms at the preceding level or is returned to the environment as heat or in parts of an organism that are not consumed by other organisms, for example, bones, teeth, hooves, and bark.

DID YOU KNOW?

Crops provide more energy for humans than cattle at a higher trophic level.

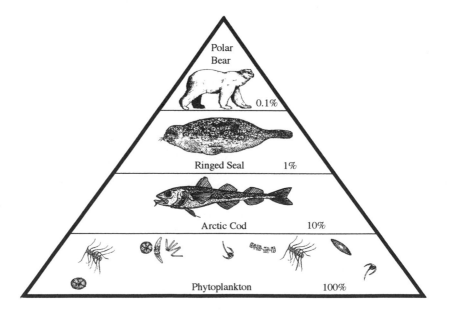

FIGURE 3-8 • Ecological Pyramid This is an example of an ecological pyramid. Notice that the greatest amount of energy in the pyramid is at the lowest trophic level—phytoplankton. Less energy is available to consumers at each higher trophic level.

Polar Bear   0.1%

Ringed Seal   1%

Arctic Cod   10%

Phytoplankton   100%

Other Ecological Pyramids

Ecological pyramids also are used to show the amount of *biomass* or the number of organisms present at each trophic level. The pyramid of numbers shows the number of organisms at each level of the food chain or food web. A pyramid of biomass displays the total mass (weight) of the organisms at each level of the pyramid of numbers.

MATTER FLOWS THROUGH ECOSYSTEMS

Organisms get their energy indirectly or directly from the sun, but they get their matter from Earth. Matter is the abiotic part of the ecosystem. It includes chemical elements and compounds such as oxygen, carbon dioxide, calcium, and nitrogen, and physical elements such as water, soil, fire, and minerals. Energy moves through ecosystems and food webs in one direction only and finally returns to the environment in the form of heat. However, the movement of matter in the environment is different. It is cyclic or circular. Unlike energy, matter remains in the ecosystem in one form or another and is used over and over again. On the other hand, energy cannot be recycled.

Water Cycle or Hydrological Cycle

The chemical properties of water are essential to living organisms. About 65 percent of the body weight of human beings is water. Without water, life cannot exist.

The water cycle shows the movement of water through the ecosystems. The water cycle is the natural circulation of water in the form of a solid, liquid, or gas in the biosphere. The water cycle circulates Earth's waters by redistributing the water that has evaporated from oceans, lakes, rivers, glaciers, ice caps, and land surfaces. The sun's radiation drives the water cycle. Solar energy causes the water in oceans, lakes,

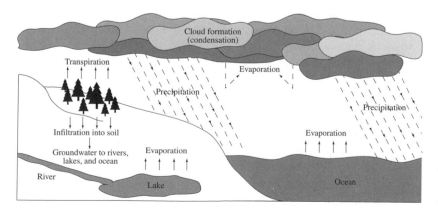

FIGURE 3-9 • Water Cycle
The hydrologic cycle or water cycle shows the natural pathway through which water moves between Earth's surface and atmosphere.

Water Facts

Why is water so important? Here are some facts.

- One human being drinks about 60,000 liters of water in an average lifetime.

- Water is the only substance on Earth that is present in the form of a solid, liquid, or gas.

- Living things are made up mostly of water. Many living organisms contain almost 66 percent water. A tomato is about 95 percent water. A chicken is about 75 percent water. A human being is about 65 percent water.

- About 70 percent of Earth's surface is covered by water. Only 3 percent of the water is fresh, and most of that is stored as ice in the polar regions.

- Water is a solvent. It can dissolve more substances than any other material.

- Because water has such a high heat capacity, it helps living things maintain a constant internal temperature and resist overheating. Water helps prevent organisms from overheating and freezing.

streams, ice caps, glaciers, and ponds to evaporate or to change from a liquid to a gas (water vapor). More than 85 percent of the water that evaporates from Earth's surface comes from the ocean. Water vapor also enters the atmosphere by transpiration, a process by which plants give off water vapor into the atmosphere. The water vapor rises into the atmosphere and when it cools, it condenses and changes into tiny water droplets forming clouds. If the droplets are heavy enough, they will fall to Earth's surface as precipitation in the form of rain, snow, hail, rain, and fog. About 80 percent of the precipitation enters the oceans. Thus there is a cycle of water movement from the atmosphere to Earth's surface and oceans and then back to the atmosphere. Eventually, all water taken in by organisms returns to the nonliving world where the water is available again for living organisms.

Carbon Cycle

Carbon is a major component of all living organisms. It is found in living organisms as well as in soil, water, and air. Like water, carbon is constantly being passed between the various parts of the biosphere.

The carbon cycle is the natural process in which carbon is cycled within the environment. Carbon is the primary component of all living matter. A naturally occurring element, like water, nitrogen, oxygen, and phosphorus, it is an essential nutrient for living organisms. Like these other substances, carbon is continuously cycled back and forth between organisms and the environment. This never-ending, cyclical process is known as the carbon cycle.

Carbon in the environment exists primarily as carbon dioxide (CO_2) gas in the atmosphere and oceans. Green plants, algae, cyanobacteria, and other producers absorb carbon dioxide from the air and water for the process of photosynthesis. During photosynthesis,

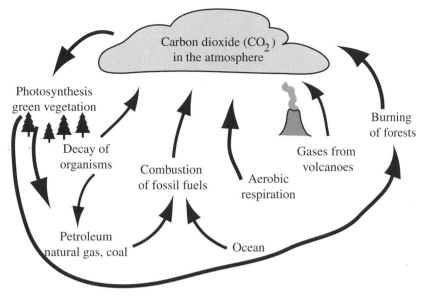

FIGURE 3-10 • Carbon Cycle The carbon cycle is a natural process in which carbon is cycled within the biosphere. Other than the water cycle, no mechanism in nature is more crucial than the circulation of carbon between the atmosphere, the lithosphere, and the hydrosphere. The human body contains about 18 percent carbon.

carbon dioxide, water, and sunlight react to form energy-rich sugars. Producers store some of the sugars; the rest provide energy, which the organisms use to carry out their biological activities. Producers also make carbon available to other organisms in the biological community. Animals and other consumers obtain carbon when they feed upon producers or other consumers. Decomposers use energy from the dead material to support their lives. In the process, the decomposers release the stored carbon in the material into the atmosphere in the form of carbon dioxide. In this way, carbon is passed along the food chain.

Carbon returns to the environment during respiration, the energy-making process in cells. Humans are living organisms which contribute to the carbon cycle by eating plants and animals and carrying out cellular respiration. During cellular respiration, the carbon-containing food molecules are broken down inside the cells to make the energy that organisms need to carry out their life processes. One waste product of this chemical reaction is carbon dioxide gas, which is released by the organisms and returned to the air, water, and soil. Scientists estimate that about 10 percent of the total amount of carbon dioxide in the air cycles back and forth between the atmosphere and organisms each year through photosynthesis and cellular respiration.

The oceans are a major storage place for atmospheric carbon dioxide. The storage place is also referred to as a sink. The ocean contains 45 times more carbon dioxide than does the atmosphere. If the oceans were to disappear, the sink would be empty. The carbon dioxide in the atmosphere would increase to a level that would cause oxygen-breathing organisms to become extinct.

Carbon is also stored in the remains of organisms when animals or plants fail to decompose completely. Fossil fuels such as coal, petroleum, and natural gas were formed over millions of years when plants become buried in sediments. Gradually, pressure and heat transform the plants into carbon-rich fossil fuels. When these fuels are burned, the stored-up carbon is released into the atmosphere as carbon dioxide.

Nitrogen Cycle

Nitrogen, another abundant chemical in Earth's atmosphere (NO_2), comprises about 78 percent of all atmospheric gases. Nitrogen, which is present in all organisms, is essential to many life processes, including photosynthesis, growth, and reproduction. Nitrogen is also a critical component of *deoxyribonucleic acid (DNA)*, the complex molecule of nuclei acids and proteins that controls cell processes in all organisms. Except for a few species of bacteria, living things cannot use nitrogen in its gaseous form.

To enter living systems, gaseous nitrogen must first be fixed (combined with oxygen or hydrogen) into compounds that can be utilized by plants, such as nitrate (NO_3) or ammonia (NH_3). The high energies provided by lightning and cosmic radiation convert some atmospheric nitrogen into nitrates. However, most nitrogen fixation is performed by species of bacteria which live in soil or in root nodules of leguminous plants, such as peas and beans. Cyanobacteria (blue-green bacteria) is able to convert nitrogen from the air into nitrogen compounds, or "fix" the nitrogen so that it is available for the plants. It is

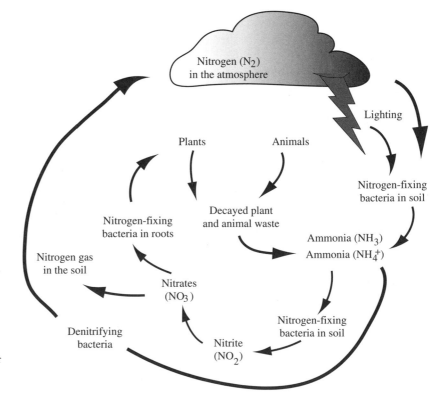

FIGURE 3-11 • Nitrogen Cycle This model illustrates the nitrogen cycle. It shows the conversion of nitrogen from one form to another through different processes.

estimated that cyanobacteria can fix as much as 100 kilograms of nitrogen per hectare (90 pounds of nitrogen per acre).

Nitrogen that has been fixed as ammonia or nitrates can be taken up directly by plants and incorporated into tissues as proteins. Animals and other consumers obtain their nitrogen by eating plants or plant-eating animals. When living things produce wastes or die, nitrogen in the form of ammonia is again returned to the soil. Some of this ammonia is taken up by plants. The rest is converted by bacteria into nitrates and nitrites through the process of *nitrification*. Nitrates may be stored in humus. Humus is the rich, organic material in soil which makes it dark. Nitrates can also be eroded from the soil and carried by moving water to rivers and lakes. They may also be converted to nitrogen gas by denitrifying bacteria and returned to the atmosphere.

The nitrogen cycle is a model or diagram that illustrates how nitrogen passes through the ecosystem. The diagram presents a flow pattern that shows how nitrogen is converted from a solid, liquid, or gas to another form through a combination of biological, chemical, or geological processes.

Phosphorus Cycle

The phosphorus cycle is the natural circulation of phosphorus in various forms through the abiotic (nonliving) and biotic (living) parts of the ecosystem. The chemical element phosphorus is indispensable to life on Earth. It is directly involved in cellular energy transfer and in the formation of DNA and various cell structures. Much of the phosphorus on Earth is tied up in sedimentary deposits and a number of phosphate-bearing rocks. In the phosphorus cycle, the chemical cycles

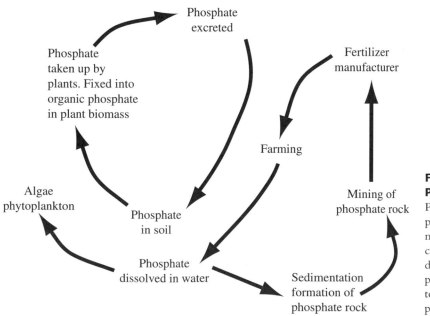

FIGURE 3-12 •
Phosphorus Cycle
Phosphorus forms phosphates a very large number of chemical compounds. The model displays the conversion of phosphates from one form to another during different processes.

continuously flow back and forth between the biotic components of the environment, such as plants and animals, and the abiotic components such as air, rocks, and water.

The phosphorus cycle, along with the iron cycle and the calcium cycle, is an example of a sedimentary cycle. Sedimentary cycles vary somewhat, but each cycle consists fundamentally of a solution phase and a rock or sediment phase. The phosphorus cycle begins when weathering and erosion break down phosphate-containing rocks. When phosphates dissolve in water, they form a solution that becomes available for algae and terrestrial green plants. Phosphorus is then passed on through the food chain as animals eat plants or other animals. When organisms decompose or excrete wastes, phosphorus is once again released into the environment for recycling.

Calcium Cycle

The calcium cycle is the natural circulation of calcium through the ecosystems. The chemical element calcium is indispensable to life on Earth. Calcium is perhaps best known for its role, along with phosphorus, in the formation and structure of bones and teeth. It also plays a role in several other vital functions in the body, including nerve transmission, muscle function, and energy production. In plants, calcium is important for growth and in the formation of cell walls and other cell structures.

Much of the calcium on Earth is tied up in sedimentary deposits of calcium carbonate, or limestone, as well as various other calcium-rich rocks. In the calcium cycle, the element cycles continuously back and forth between the biotic and the abiotic components of the environment. The calcium cycle occurs when weathering and erosion break down limestone and other calcium-containing rocks. Limestone often dissolves very easily in water. When this happens, a solution is formed, making calcium available for algae and terrestrial green plants. Calcium is then passed on through the food chain as animals eat plants or other animals. When organisms die and decompose, calcium is once again released into the environment for recycling.

Vocabulary

Biomass The total mass of living organisms present in a given area.

Carbohydrate A class of organic compounds composed of carbon, oxygen, and hydrogen, including sugars and starches.

Deoxyribonucleic acid (DNA) An organic compound that carries the genetic or hereditary information for virtually all organisms. As an example, genetic information is transmitted from parent to offspring so that the offspring looks like their parents. RNA (ribonucleic acid) is a related chemical that transmits information from DNA to form enzymes and proteins.

Energy The ability to do work. Light, heat, electricity, and sound are forms of energy. Energy normally is not created or destroyed but may be changed from one form to another. As an example, electrical energy can be changed to heat or light.

Matter Anything that has mass and takes up space.

Nitrification The process in the nitrogen cycle that results from the action of nitrifying bacteria using oxygen to change oxidized ammonia in wastewater into nitrites and then to nitrates.

Pigment A colored chemical which is found in the body parts of plants and animals. Skin, feathers, fur, leaves, and flowers contain pigments.

Prey An animal that is killed and consumed by another animal.

Trophic level A feeding step in a food chain.

Activities for Students

1. Draw a food chain including human beings. On which trophic level are humans? What animals are at more than one trophic level?

2. Consider the water cycle. Is water a renewable or nonrenewable resource? Once water is used in human activities, how is it funneled back into the water cycle? Learn more about a water-treatment plant in your area.

3. Plants produce energy by absorbing carbon dioxide. Why is this beneficial to humans? If more plants are destroyed, what will happen to carbon dioxide levels in the atmosphere? How will this subsequently affect human life?

Books and Other Reading Materials

Capula, Massimo. *Simon & Schuster's Guide to Amphibians and Reptiles of the World*. New York: Simon & Schuster, 1989.

Cogger, Harold G., and Richard G. Zweifel, eds. *Encyclopedia of Reptiles & Amphibians*. San Diego, Calif.: Academic Press, 1998.

Kirk, John T. O. *Light and Photosynthesis in Aquatic Ecosystems*. Cambridge, England: Cambridge University Press, 1994.

Lauber, Patricia. *Who Eats What? Food Chains and Food Webs*. Illustrated by Holly Keller. New York: Harper-Collins Children's Books, 1995.

Wallace, Holly, and Anita Ganeri. *Food Chains and Webs (Life Processes)*. Boston: Heinemann Library, 2001.

Websites

Energy.Government, Kidzzone, environmental resources for projects, http://www.energy.gov/kidz/sub/sciencelinks

Geographic Information Service, GIS Software Programs for Classroom project, http://metroeast_climate.ciesin.columbia.edu/gis

Internet Resources, Environmental Education, World Resources Institute, educational resources, http://www.igc.org/wri/enved/edulinks

U.S. Geological Survey, information on the carbon cycle in lakes in the Upper Mississippi River Basin (Wisconsin, Minnesota, and North and South Dakota), http://climchange.cr.usgs.gov/info/carbon/

U.S. Geological Survey, United States Global Change Research Program: Carbon Cycle Science Program, http://geochange.er.usgs.gov/usgcrp/ccsp/

CHAPTER 4

Land Biomes: Forests

Forests are large, global ecosystems or biomes which have been growing naturally for hundreds and hundreds of years. They are one of Earth's most important natural, *renewable* resources. Human societies, past and present, have benefited from the ecological, aesthetic, and economic contributions of forests.

The world's forests occupy about 30 percent of Earth's surface. They support a larger number of plant and animal species than any other ecological system. Forests play a major role in the recycling of carbon, nitrogen, and oxygen. Trees in the forests maintain soil fertility, control soil erosion, protect watersheds, and provide recreational areas. Approximately 50 percent of the world's population depends on forests for fuelwood. Among the many goods supplied by forests, the principal ones are lumber products including many building materials such as planks, plywood, hardboard, and chipboard as well as paper products and paperboard. Other forest products include medicines; fruits, nuts, and other foods; spices; rubber; and gums.

There are many different types of forests. Some of the major ones include the coniferous forest, temperate *deciduous* forest, temperate rain forest, tropical rain forest, and the urban forest.

TAIGA FORESTS

A coniferous forest is a forest biome in which most of the trees are conifers. Conifers are needle-leafed, cone-bearing, evergreen trees such as spruce, cedar, pine, fir, and hemlock. Most conifers lose only a

FIGURE 4-1 • Forest Products

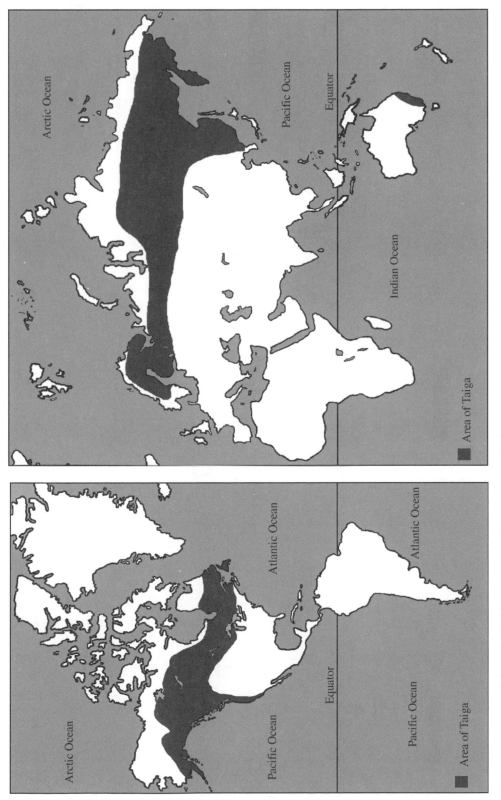

Area of Taiga

FIGURE 4-2 • Areas of the Taiga in the Eastern and Western Hemispheres

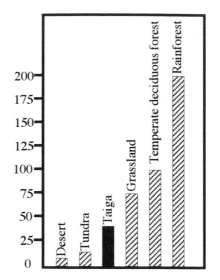

FIGURE 4-3 • Average Annual Rainfall of Biomes (cm/year)

FIGURE 4-4 • Cones are part of the coniferous tree where the seeds are found. There are about 550 species of conifers.

few of their leaves at a time. Needles stay on the tree until new needles grow. In fact, the needles on conifers can last for several years before dropping off. The cones are the part of the tree in which the seeds are found. The scales of the cone protect seeds until the cone ripens and releases its seeds.

As shown in Figure 4-2, the coniferous forest biome occupies about 35 percent of the world's forested land. It extends in altitude from *alpine tundra* to temperate deciduous forests to subtropical regions. The Alaskan interior, the southeastern United States, the Rocky Mountains, Canada, northern Asia, Russia, and northern Europe all have coniferous forests. The largest forested area covered by the coniferous biome is known as the taiga.

The taiga, also known as boreal forest and northern coniferous forest, is the northernmost coniferous forest. The taiga is located in a nearly continuous belt in northern Eurasia, northern Europe, Siberia,

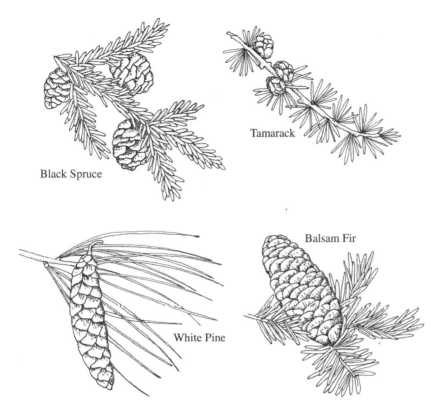

Black Spruce

Tamarack

White Pine

Balsam Fir

FIGURE 4-5 • Some Needle-leaf Trees Species of conifers can be identified by the characteristics of their needle-like or scaly leaves.

northern Japan, northern Alaska, and Canada. The taiga belt is between 45° and 57° north latitudes. The northern edge of the taiga is defined by a treeline south of the Arctic tundra. Very cold winters, high soil moisture, densely packed trees, and abundant snow characterize the taiga coniferous forest biome. The biome is covered with lakes and ponds, which eventually fill with mineral and organic matter to form bogs.

Taiga Forest Ecosystem

In the taiga forest, most tree species fall into one of four categories: spruce, fir, pine, and tamarack or larch. There is a short growing season of around 130 days in the taiga, with a wide annual temperature range. The trees grow to a height of from 10 to 18 meters (30 to 50 feet) and have diameters of from 3 to 5 centimeters (8 to 12 inches) at maturity. If taiga conifers are cleared by fire or logging, broad-leafed species, such as aspen and poplar, will begin to recolonize the land because the new soil conditions are suitable for these kinds of trees.

Taiga surface soils are acidic because of the acidic substances in the buildup of fallen needles on the forest floor. The soil is also deficient in nutrients because minerals are leached or washed out of the soil by rainwater. As a result, the dark, humid floor is suitable only for a few plants and fungi which require little light. Below the surface soil is *permafrost*, which exists in about 65 percent of the taiga's range. The

DID YOU KNOW?

Not all needle-leaf trees are evergreen. A few of them shed all of their leaves during the fall season of each year. One of these evergreens is the tamarack or larch, which is found in the taiga forest. The tamarack is called a deciduous conifer.

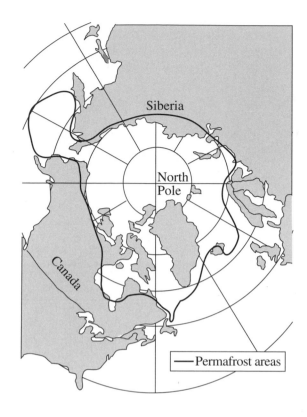

FIGURE 4-6 • An estimated 20 to 26 percent of Earth's land area is covered by permafrost that likely formed during the last ice age.

plants on the forest floor include ferns and club moss. Lichens and mosses are found on most trees.

Animals in this forest include herbivores such as moose, deer, snowshoe hares, caribous, elks, beavers, squirrels, and mice. Near taiga streams, beavers make their homes in the forest. They cut down trees and build dams of tree branches and twigs. The dams trap the water providing homes for amphibians, waterfowl, and fish. Wetlands within the taiga are breeding grounds for a variety of ducks and swans. Other taiga birds include migratory insect eaters, such as wood warblers, and year-round seedeaters, such as finches. Predators include bears, wolves, lynx, bobcats, wolverines, foxes, and predatory birds such as hawks, owls, and eagles. Bark beetles, wood wasps, springtails, and other invertebrates—those without backbones—live on the foliage, buds, needles, and bark of coniferous trees.

Environmental Concerns of Taiga Forests

Refer to Volume IV for more information about environmental concerns of forests.

The taiga has been extensively exploited for timber in the southern portions of its range; most American softwood lumber comes from this type of forest. Other disruptions in the taiga ecosystem have included overhunting and trapping, which have reduced the numbers of fur-bearing animals, such as mink and marten. Mining activities remove

Red squirrel

Lynx

Wolf

Goshawk

Short-tailed weasel

Moose

Figure 4-7 • Taiga Animals The Taiga biome lies south of the tundra and is characterized by a coniferous forest composes of pine, fir, hemlock, and spruce.

surface soil and vegetation destroying habitats and causing soil erosion and pollution from mining wastes.

TEMPERATE FORESTS

The two kinds of temperate forests include temperate deciduous forests and temperate rain forests. Temperate describes a region or climate which is neither very hot nor very cold. The Temperate Zone is

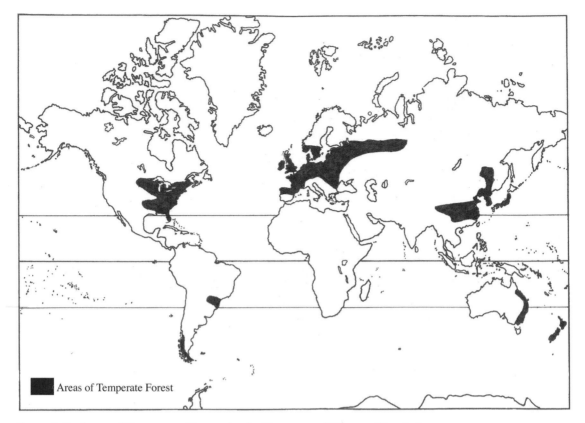

FIGURE 4-8 • **Areas of Temperate Forests in the Eastern and Western Hemispheres**

found between the polar regions and the tropics, primarily between 30° and 50° north latitude. Temperate forests are also referred to as midlatitude deciduous forests.

Temperate Deciduous Forests

A deciduous forest is a forest made up primarily of deciduous trees. The dominant plants in this forest are broad-leafed, thick-barked, deciduous species including maple, beech, oak, hickory, and elm. Because the biome covers large geographical areas, it is generally subdivided into forest regions on the basis of the domination of a particular tree species, such as beech, or association of species, such as a beech-maple forest. Some conifers, mostly pines, are usually mixed in with hardwoods, which are slow-growing broad-leafed trees such as maple and oak.

Temperate deciduous forests are located in parts of North America, central Europe, and northeast Asia in the countries of Japan, Korea, and China. Much of the eastern half of the United States and southeastern Canada are considered to be temperate forests. Average precipitation ranges from about 75 to 150 centimeters (30 to 60 inches) per year. Deciduous forests have hot summers and cold winters. Temperatures can reach as high as 35°C (95°F) during the summer months and dip well below freezing in the winter. Frost occurs throughout the biome.

FIGURE 4-9 • The Goshawk lives in a temperate deciduous forest, which is a Northern Hemisphere biome, characterized by moderate climate, relatively high rainfall, well-defined seasons, and mostly deciduous trees.

FIGURE 4-10 • Oak (left) and maple leaves are found in deciduous forests. Deciduous plants lose all their leaves at the same time.

DECIDUOUS FOREST PLANTS

Temperate deciduous forests are characterized by seasonal changes in temperature. Most of the trees shed their leaves for several months each year. Deciduous trees in cold regions shed their leaves in the fall in advance of winter drought to prevent frost injury. Leaves change color and drop in fall because shortening days cause trees to withdraw chlorophyll from their leaves and halt photosynthesis in preparation for winter dormancy. Many deciduous trees will also shed their leaves during excessive hot weather to avoid drought injury.

American Beech

Sugar Maple

Birds

White-tailed Deer

Moose

Black Bear

Beaver

Racoon

Waterfowl

Grasses

Ferns

Fish

Figure 4-11 • Deciduous Forest Ecosystem Many plants and animals live in a deciduous forest ecosystem.

In a mature deciduous forest, the *canopy* trees grow from 20 to 35 meters (60 to 100 feet) high. An *understory* of saplings, annual and perennial plants, and herbs is typically well developed. Bushes, shrubs, and particularly small species flower in the spring before the leaves of the tree canopy grow and block out the sunlight. The growing season in temperate deciduous forests ranges from 140 to 300 days per year, and the average temperature is 10°C (about 50°F).

The soil of the forest floor consists of a top layer of humus and a deeper layer of clayey soil. One hectare (2.5 acres) of good-quality soil in a temperate zone contains hundreds of millions of invertebrates, including mites, millipedes, and worms. Herbs, ferns, small shrubs, and wildflowers occupy the ground. Fire can be a positive factor in maintaining forest health and the diversity of herbs and shrubs on the forest floor.

DECIDUOUS FOREST ANIMALS

Temperate deciduous forest fauna include black bears, brown bears, white-tailed deer, foxes, chipmunks, moose, raccoons, skunks, beavers, and resident and migratory birds. Large carnivores have been largely eliminated through human hunting, but do include animals such as timber wolves, cougars, and bobcats. The coyote has moved eastward and taken over the timber wolf's niche in many temperate deciduous forests. Predatory birds include hawks, owls, and eagles. Snakes, lizards, frogs, and salamanders are also common inhabitants.

ENVIRONMENTAL CONCERNS OF TEMPERATE FORESTS

This biome has been very extensively affected by human activity. Much of it has been converted into agricultural fields, which has resulted in a significant loss of *biodiversity*. Other detrimental impacts

Rain forests exist in different countries besides Brazil and Peru. This rain forest is located in British Columbia, Canada. (Courtesy of Gary Friedman)

have resulted from timber harvesting, fragmentation by highways and rights of way, pesticide use, acid rain, air pollution, and suppression of natural fires.

Temperate Rain Forests

This primarily coastal, coniferous forest biome is characterized by thick stands of trees, high rainfall, and high humidity. This ecosystem, also known as a moist coniferous forest, occurs in wet, cool climates where the ocean air meets coastal mountains.

Though fewer in number than in the tropics, rain forests grow in temperate regions where conditions favor their development. Temperate rain forests are located in areas between 32° and 60° degrees latitude with at least 200 centimeters (80 inches) of annual rainfall. Temperate rain forests have a more seasonal climate than do tropical rain forests near the equator, with less constant temperatures and less rain.

Temperate rain forests have always had very limited worldwide distribution. Today, undisturbed forests of this type exist in Chile and in the Pacific Northwest region of North America, where they extend in a narrow band from the southern part of Alaska to central California. There are also some temperate rain forests on the western slopes of the Rocky Mountains in Idaho and British Columbia.

PRECIPITATION AND SOIL

Abundant precipitation is characteristic, in the forms of both rain and snow. Fog is also an important component in the southern portion of the range. Cool temperatures are usual, with only a moderate range from summer to winter. In most places, the forest is thick, with a dark floor, and understory species are limited to mosses and ferns because of the lack of light.

Soils in the temperate rain forest are deep and rich, but mineral poor. They support diverse conifers and hardwoods, many of which are very tall and old. Redwoods, for example, have thick, insect- and disease-resistant bark, which helps some individual trees live to 1,000 or 2,000 years. Many native species of shrubs and wildflowers grow, and the forest provides habitat for numerous resident birds and mammals.

Fur-bearing mammals, such as otters, were once common before being overhunted since the seventeenth century. However, temperate rain forests typically contain many types of mammals

The Olympic Rain Forest located in Washington State is one of the world's largest and most impressive coniferous temperate rain forests. Although the Olympic Rain Forest is located between 50° and 60° north latitude, the forest never freezes because the Pacific Ocean helps regulate temperatures. In this area, winds travel from west to east and air picks up a lot of moisture from the Pacific. As the air moves up the western side of the Olympic Mountains, its moisture is wrung out over the coastal forest. Up to 415 centimeters (167 inches) of rain falls annually in the Olympic Rain Forest, which receives more rainfall than any other place in the continental United States.

and birds, including woodpeckers, hawks, owls, squirrels, shrews, moose, deer, and wolves.

TEMPERATE RAIN FOREST ECOSYSTEM

Though plant life and animal life are abundant, the species are not as diverse as they are in the warmer, tropical rain forests. Towering evergreen trees such as Douglas fir and Sitka spruce, some of which are more than 1,000 years old, dominate the forest. Mosses, *lichens*, and epiphytes

FIGURE 4-12 • A classic old growth forest contains giant redwoods (above), cedars, Douglas fir, hemlock, or spruce.

(plants that live on other plants for physical support) drape the tree branches and cover massive tree trunks. The damp, dark, forest floor is literally carpeted with ferns, fallen and decaying logs, mushrooms, and new seedlings. Insects, including mosquitoes and flies, are common inhabitants of temperate rain forests. Few cold-blooded animals, such as frogs, salamanders, turtles, and snakes, live in these forests because of the low temperatures.

U.S. Forest Service

The U.S. Forest Service is the largest agency of the U.S. Department of Agriculture. Congress established the Forest Service in 1905 to provide water and timber for the nation's benefit. Congress directed the agency to manage national forests for additional multiple uses and benefits, and for the sustained yield of renewable resources such as water, wildlife, and recreation. The Forest Service's mission is summarized as "Caring for the Land and Serving People." The agency manages natural resources under the best combination of uses to benefit the American people, ensure the productivity of the land, and protect the environment. To find your National Forest by state, refer to http://www.fs.fed.us/recreation/states/us.shtml.

ENVIRONMENTAL CONCERNS OF TEMPERATE RAIN FORESTS

Commercial timber activities have dramatically reduced the extent of this forest type as a result of earlier clear-cutting operations which removed all of the vegetation. Clear-cutting is generally no longer practiced. In recent years, fire-suppression efforts have increased tree losses by enabling the buildup of fuel for intense forest fires.

TROPICAL RAIN FORESTS

Tropical rain forests are found along the equator extending 10° north and 10° south and at various altitudes. One can find rain forests in lowland swamps and in warm alpine regions. These vast tropical forests are located in South America, Africa, and southeast Asia. The largest rain forests are found in Brazil (South America), Zaire (Africa), and Indonesia (southeast Asia). Brazil, however, has one-third of the world's tropical rain forest. Other tropical rain forests are located in Hawaii and the islands of the Pacific and Caribbean oceans.

Tropical rain forests are especially fertile ecosystems. The combination of steady heat and moisture creates an environment that encourages growth of many kinds of plants. As a result, tropical rain forests have a greater amount of biodiversity than other land biomes. As an example, a hectare (2.5 acres) of a tropical rain forest in South America may have 300 tree species while a hectare of forest in North America may have 30 tree species. This example illustrates that biodiversity tends to increase as one moves toward the equator. Biologists estimate that two-thirds of the world's plant species grow in tropical rain forests today, and about 70 percent of all plant species in these forests are trees.

On an average, tropical rain forests receive from 150 to 400 centimeters (55 to 160 inches) of rain every year. The temperature ranges from 25° to 35°C (77° to 95°F). With few significant seasonal changes—no wide fluctuations in temperature, no long, dry spells—most rain forests essentially have an endless year-round growing

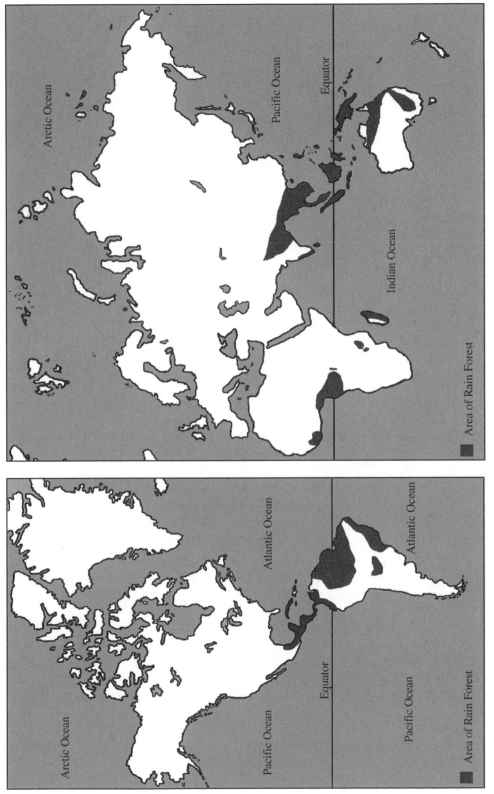

FIGURE 4-13 • Areas of Rainforest in the Eastern and Western Hemispheres

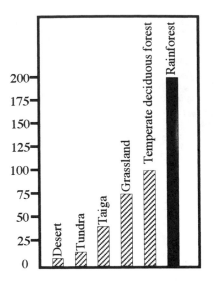

FIGURE 4-14 • Average Annual Rainfall of Biomes (cm/year)

season. This condition creates a great food supply which can support a large variety of organisms.

Rain Forest Ecosystem

A tropical rain forest has three layers: the canopy, the understory, and the forest floor. These layers, however, blend together and their differences allow many organisms to find a niche.

The canopy is made up of treetops, which may rise from 15 to 45 meters (50 to 150 feet) above the ground. Most plants, such as palm trees, have large, broad leaves in order to maximize capture of the ample sunlight. Other trees include mahogany, teak, and ebony.

Most of the animals in the rain forest, such as monkeys, birds, tree frogs, and even snakes, live in the canopy. The canopy contains an entangled maze of woody vines called lianas. It also contains a rich epiphyte community, made up of orchids and bromeliads, which grow on trunks and branches of trees to access the sunlight. Very little sunlight, as low as 2 percent, filters down through the canopy. Thus the understory consists of young trees, dwarf palms, ferns, and broad-leafed shrubs which never grow to adult size.

The forest floor, or the bottom layer of the rain forest, is where decomposition occurs rapidly in the warm, damp climate. Earthworms, termites, ants, and fungi are busy breaking down organic matter. Except for rotting vegetation, which nourishes the thin tropical soil, the forest floor is almost bare.

Animal Species

The abundant vegetation of the rain forest provides food and shelter for millions of animal species. Most of this animal diversity is made up of insects, but many other animal groups are also represented. It is

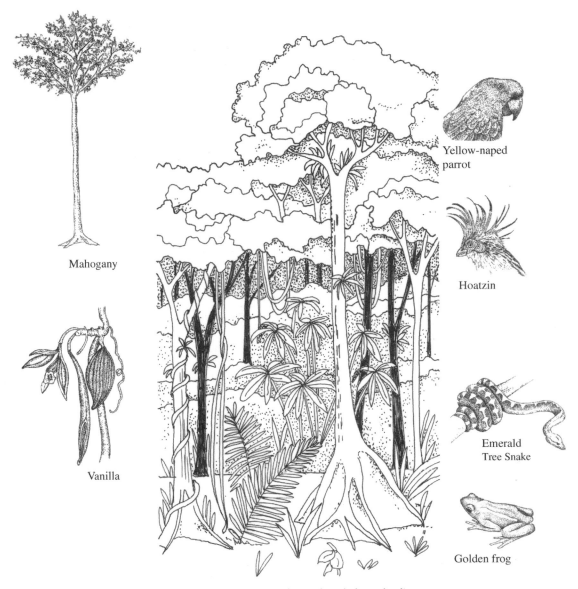

Mahogany

Vanilla

Yellow-naped parrot

Hoatzin

Emerald
Tree Snake

Golden frog

FIGURE 4-15 • The tropical rain forest has many species of animals and plants that live at different layers of the forest. A hectare of tropical rainforest in Amazonian Peru may have 300 tree species. The layers in the rainforest include the canopy, the understory, and the forest floor.

estimated that 1,000 hectares (2,500 acres) of rain forest in South America contains thousands of insect species, including 150 different kinds of butterflies. One study identified 600 different species of beetles in the canopy of just one tree.

The forest also contains dozens of species of poisonous snakes and frogs. One kind of poison frog is the poison arrow frog found in the Central and South American tropical forests. South American indigenous peoples used the poison from these frogs on the tips of

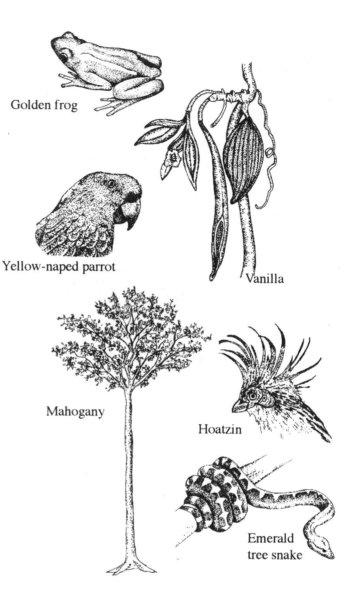

Golden frog

Yellow-naped parrot

Vanilla

Mahogany

Hoatzin

Emerald tree snake

FIGURE 4-16 • Some animals and plants that live in a tropical rain forest.

their arrows and blowpipe darts. The toxic chemicals from one species, the dyeing poison arrow frog (*Dendrobates tinctorius*), are so harmful that just holding this frog in your hand can be dangerous.

There are hundreds of varieties of brightly colored birds, such as toucans, parrots, and hummingbirds. Mammals include the tree-dwelling sloths, monkeys, and fruit bats, as well as jaguars, tapirs, ocelots, and gorillas which prowl the forest floor.

Mutualistic Interactions

Refer to Chapter 2 for more information about mutualism.

Mutualism takes place in the rain forest. For example, many animal groups, especially insects and birds, pollinate rain forest trees. The insects receive food from the nectar, and in return they pollinate more flowers.

The Penan hunters are indigenous people who live in the Brazil rain forest. (Courtesy of Rainforest Action Network)

After the fruit is formed, rain forest plants also rely on animals to disperse their seeds. Birds and mammals are important dispersal agents in nearly all rain forests. Animal groups can also provide protection to a plant species, while the host plants provide the animals with a home. Many species of stinging and biting ants, for instance, live in the hollow stems of tropical plants. The ants supply nutrients to the trees and in many cases also protect the trees from leaf and seed predators.

Environmental Concerns of Tropical Rain Forests

Many natural disasters have had an impact on rain forests. Cyclones, forest fires, disease, and landslides are natural forces that have all impacted rain forest ecosystems. The environmental damage caused by these phenomena, however, is rather minimal compared to such human activities as logging, road building, mining, and large-scale deforestation.

Scientists are concerned because the rate of rain forest loss far exceeds the rate of growth. At an earlier time, tropical rain forests once covered more than 1.6 billion hectares (4 billion acres) of Earth. Today, nearly half the tropical rain forests are gone. They now cover only about 7 percent of the land. The large-scale deforestation in nearly all rain forest areas today is so extensive that it would take hundreds of years for anything resembling the original vegetation to return. When hectares of trees are lost, countless animals are driven from their homes. Animal and plant populations have declined with habitat destruction.

Slash and burn is a forestry practice that is used in tropical rain forests to clear the land for farming and cattle raising. (Courtesy of Rainforest Action Network)

Rain forests contain Earth's greatest diversity of plants and animals; they also represent giant gene banks that can provide new drugs, foods, and other products for people. Medicinal substances already discovered in rain forests include reserpine, used to treat cardiac problems, and curare, used in heart and lung surgery. Scientists fear that rain forests will disappear before they can learn what other medicines can be found.

Rain forest trees also prevent soil erosion. The roots bind the soil together and help prevent runoff of the topsoil.

Governments, scientific organizations, conservationists, and other citizens are deeply concerned about the loss of rain forests. The future of Earth's rain forests may depend on management plans that preserve some areas of rain forest while allowing people to cut trees selectively in other areas.

Better land-use practices, education, and wiser planning may slow deforestation, but experts worry that the rain forests will be gone by the time these changes can be widely implemented.

Refer to Volume II to learn more about the medicinal resources in rain forests.

URBAN FORESTS

The nation's forested lands in and adjacent to urban areas is urban forest. Urban and community forests are made up of trees and associated woody vegetation within the surroundings of populated places. The urban forest encompasses trees along streets, in greenbelts, and in city parks, municipal watersheds, and similar areas. There are 29 million hectares (72 million acres) of such vegetation in the United States; 80 percent of Americans live or work among them daily.

Urban and community forests provide many benefits, including the reduction of energy costs through summer shading and winter wind protection. Urban areas have distinctly different temperature

Old-Growth Forests

A classic *old-growth forest* contains giant redwoods, cedars, Douglas fir, hemlock, or spruce. It has at least 20 large trees per hectare (2.5 acres) which are more than 300 years old, or are more than a meter in diameter at breast height. In Washington's Mount Rainier National Park, for example, many trees are between 500 and 1,000 years old. There are hundreds of species of ferns, vines, mosses, lichens, shrubs, and understory trees beneath the towering canopy trees.

The Pacific Northwest Rain Forest is perhaps the most well-known old-growth forest. Here there are dense, closed canopies of redwood or Douglas fir, a fern-covered forest floor, and large moss-covered dead trees decaying on the ground. In the arid West, by comparison, ponderosa pines grow to large sizes in relatively open conditions; southern pine forests exhibit similar characteristics. These two ecosystems often are dependent on frequent, light-intensity ground fires to thin out competing vegetation. In the northern Rocky Mountains, old hemlock forests are dominated by dense hemlock canopies with little vegetation underneath except for hemlock seedlings and some shade-tolerant plants, such as orchids. Alaska's Tongass National Forest is one of the world's largest tracts of temperate old-growth forest. Elsewhere in the world, old-growth forests exist in Canada, Russia, Mexico, and Central America.

The importance of the old-growth forest ecosystem extends far beyond the preservation of this one species, however. Old-growth forests provide habitat for plants and animals that cannot live anywhere else. At least 118 known vertebrate species live primarily in old-growth forests; 41 of them cannot nest, breed, or forage in any other location. In contrast, new-growth forests are managed as monocultures, containing only a few species of trees that were planted at the same time; they do not offer habitat diversity. For example, only 9 mammal species make their home in second-growth forests of young firs and hemlocks, compared to 25 species in old-growth forests containing the same tree species. Loss of forests throughout the world has led to a greater awareness of the value of the remnants of older natural forests which are relatively free of modern human disturbance.

regimes from those of large parks and rural areas; they are usually warmer, which is a condition known as the urban heat island. Summertime studies have shown a 0.5° to 1.0°C decrease in temperature for every 10 percent increase in vegetation cover. Cooling energy needs for houses shaded by trees are from 4 to 25 percent less than for houses located in open areas. Urban trees will be increasingly important in energy conservation as fossil fuels for cooling and heating become more costly and scarcer.

Randy Hayes

Randy Hayes is the founder and president of the Rainforest Action Network (RAN) in San Francisco, California. The RAN works with environmental and human rights groups around the world, including indigenous forest communities and nongovernmental organizations in rain forest countries.

The RAN's Protect-an-Acre program gives small amounts of money, called grants, to communities of indigenous people who live in the rain forest. The communities then use the money to help protect their forest homes. The Protect-an-Acre program is funded by students, classrooms, schools, and individuals who raise money to save rain forests.

Randy Hayes is an environmental activist and the president of Rainforest Action Network based in San Francisco, California. (© Rainforest Action Network)

Benefits of Urban Forests

Additional benefits of urban forests include control of stormwater runoff and erosion; filtration of airborne particulate matter and pollutants, such as ozone; consumption of carbon dioxide; production of oxygen through photosynthesis; and cooling through evapotranspiration, the loss of water in an area caused by the evaporation of water from land and transpiration from plants. Planting urban trees is a good way to address the greenhouse effect and the global warming problem.

Urban forests also directly increase property values by making communities more attractive. They offer recreational opportunities and provide urban wildlife habitat. They aid in human physical and psychological health, community stability, and crime-reduction efforts. These forested areas provide city dwellers with the opportunity to experience, understand, and appreciate forest-related benefits.

It is costly to establish and maintain urban and community forests. Larger saplings and full-grown trees, generally used more than seedlings or saplings, are expensive. In addition, there are high maintenance costs because of challenges related to utility line clearance, storm damage repair, debris removal, and disease protection. Some costs and possible problems can be reduced or avoided through proper selection and location of trees. The average life span of urban trees ranges from 7 to 10 years in the downtown busy area of a city. City trees growing in a more open area, such as a park, can live for about 50 years.

Environmental Concerns of Urban Forests

Over the past several years, shrinking city budgets have produced problems for some of the nation's urban forests. Traditionally, city governments have managed city trees; however, there have been cuts in spending for urban forest management. More and more, the responsibility has shifted toward nonprofit organizations, which have also been important in raising public awareness of and involvement in the urban forest.

Vocabulary

Alpine tundra The grassland area found above the tree line on mountain ranges.

Biodiversity The number of species that live in a certain area.

Canopy The top layer covering of tall trees in a forest.

Deciduous Species of trees that lose their leaves each year.

Lichen An organism composed of a fungus and an alga growing in a mutual relationship and forming a dual organism. They are found often on trees and rocks.

Old growth A late stage in the life of a forest's ecosystem in which there are many large mature trees.

Permafrost Permanently frozen soil below the topsoil.

Renewable A natural resource in abundant supply which can replenish itself.

Understory Trees, shrubs, and plants that live below the canopy.

Activities for Students

1. On a world map, create a color-coded visual representation of the areas where the various types of forests can be found. One hundred years ago, how might this map have looked different?

2. Tropical rain forests are shrinking at a startling rate for economic and development reasons. Brainstorm governmental and business policies that would focus on the sustainable uses of the rain forest as a natural resource.

3. Contribute to the urban forest biome in your area. Coordinate with a school or community group to plan a tree-planting activity in your area.

Books and Other Reading Materials

Breymeyer, A. I., D. O. Hall, and J. M. Melillo. *Global Change: Effects on Coniferous Forests and Grasslands*. Scope no. 56. New York: John Wiley and Sons, 1997.

Lewington, Anna. *Atlas of Rain Forests*. Atlases Series. Austin, Texas: Raintree/Steck Vaughn, 1997.

Norse, Elliot A. *Ancient Forests of the Pacific Northwest*. Sponsored by the Wilderness Society. Washington, D.C.: Island Press, 1990.

Sayre, April Pulley. *Temperate Deciduous Forest* (Exploring Earth's Biomes). New York: Twenty First Century Books, 1995.

Terborgh, John. *Diversity and the Tropical Rain Forest*. Scientific American Library no. 38. New York: W. H. Freeman, 1992.

Yahner, Richard H. *Eastern Deciduous Forest: Ecology and Wildlife Conservation*. Vol. 4 of *Wildlife Habitats*. Duluth: University of Minnesota Press, 1996.

Websites

American Conifer Society (ACS), http://www.pacificrim.net/~bydesign/acs.html

American Forests, http://www.americanforests.org

Rainforest Action Network, http://www. ran.org

Society of American Foresters, http://www. safnet.org

U.S. Forest Service, http://www.fs.fed.us

World Resources Institute Forest Frontiers Initiative, http://www.wri.org/ffi

World Wildlife Fund (Worldwide Fund for Nature) Forests for Life Campaign, http://www.panda.org/forests4life

CHAPTER 5

Other Land Biomes

The temperate and tropical forests discussed in Chapter 4 are flanked by other land biomes. These major land biomes include grasslands, deserts, chaparrals, and tundras. All of them do not receive enough precipitation to sustain a forest. The limited amount of rainfall and climate conditions result in a decrease in the diversity of species.

GRASSLANDS

Grasslands are dry, flat biomes that are dominated by grasses and shrubs, which are the *climax* vegetation. A few trees, such as cottonwood, sand plum, and hackberry, exist along the banks of streams, rivers, and marshes in some grasslands. Grasslands represent about one-third of Earth's landmass, though most have been converted to vast farmlands or livestock-grazing lands because of the rich, fertile soil and the availability of water from underground sources. Grasslands are found in many parts of the world including both North and South America, Siberia, Africa, Australia, and China.

In general, all grasslands are rather warm and dry. Average precipitation in temperate and tropical grasslands (savannas) ranges from 25 to 75 centimeters (10 to 30 inches) annually. In tropical grasslands, such as those in Africa and South America, temperatures are fairly consistent throughout the year, remaining around 30°C (86°F). In temperate grasslands, such as those in the central United States and central Europe, temperatures show great fluctuation, from about 0°C (32°F) in the winter to about 25°C (77°F) in the summer. Grasses are adapted to surviving in the sometimes drought-like conditions of grasslands. Their root systems, for instance, form thick underground mats and are able to survive long droughts and brush fires.

Precipitation in the grasslands is irregular and varies from season to season during the year. There are times of heavy rainfall and periods of dry weather. Precipitation averages between 25 to 75 centimeters (10 to 30 inches) a year. In grasslands with high precipitation, grasses can grow as high as 3 meters (9 feet) tall. In the drier regions, grasses grow between 25 to 50 centimeters (10 to 20 inches) tall.

Grasslands are habitats for a wide variety of animals and plants. Depending on the location of the grassland, there may be bison, eagles,

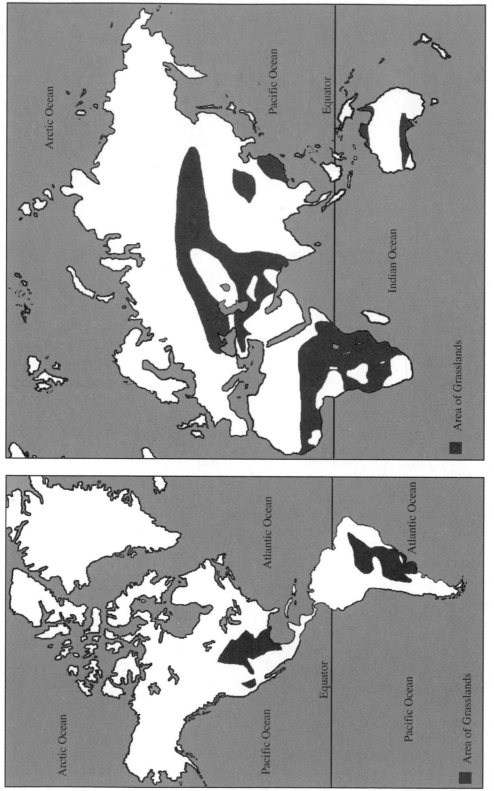

FIGURE 5-1 • Areas of Grasslands in the Eastern and Western Hemispheres

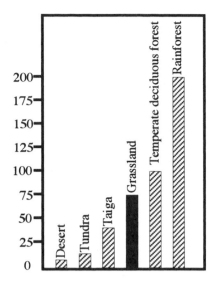

FIGURE 5-2 • Average
Annual Rainfall of
Biomes (cm/year)

deer, and bobcats, as well as cheetahs and wildebeest. Many grassland animals, such as horses, giraffes, and buffalo, are adapted to feeding on grasses. Wildebeest and zebras eat different grasses at different times of the year. The predators that play a key role in the grassland ecosystem include Eurasian jackals, African wild dogs and lions, and North American coyotes. The grasslands also serve as flyways for migrating birds such as Canada geese, white pelicans, cranes, and wild ducks.

There are two major types of grasslands: temperate grasslands and tropical grasslands. Temperate grasslands include prairies, pampas, and steppes; tropical grasslands are called savannas.

Temperate Grasslands

Temperate grasslands are found in mostly dry and flat interior sections of the continents. These grasslands are characterized by high temperatures, low precipitation, and drought-like conditions.

PRAIRIES

The grasslands of central North America, called prairies, extend from south-central Canada to Texas, and from the Rocky Mountains to Ohio. The North American prairie can be divided into three broad categories: tallgrass, short grass, and mixed.

The tallgrass prairie occurs in the easternmost section, extending into relatively fertile soil and humid regions. The name tallgrass derives from the prevalence of upright bluestem grasses, which reach heights of over 2 meters (6 to 10 feet). Today, the tallgrass prairie has been almost completely converted to cropland for agriculture.

Short-grass prairie is located in the drier, less fertile soil of the westernmost extent of the prairie's range. Vegetation is mainly short, bunch grasses ("sod grasses") which grow no more than 50 centimeters

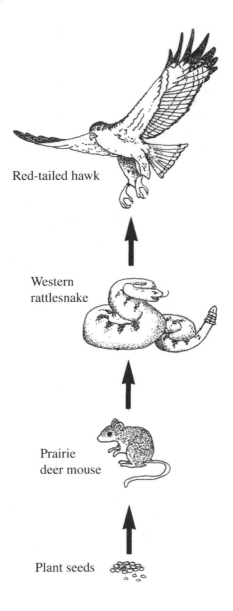

Red-tailed hawk

Western
rattlesnake

Prairie
deer mouse

FIGURE 5-3 • This is an example of a prairie food chain showing different consumers.

Plant seeds

(20 inches) high. Today, many of these grasslands are cattle ranches. A mixed-grass prairie contains grasses that grow between 50 and 150 centimeters (20 and 60 inches) tall.

Undisturbed prairie provides habitat for grazing species such as deer, wild horses, and bison (once numbering in the millions, but now limited to private or reintroduced herds). Predators include eagles, bobcats, coyotes, and black-footed ferrets (an endangered species), which feed on prairie dogs, cottontail rabbits, white-tailed deer, and squirrels. Some animals, such as prairie dogs and badgers, are able to dig extensive underground burrows to avoid detection by predators.

Frequent burning by fire has played a key role in the creation and maintenance of prairies; in fact, some grasses require periodic fires in order to grow new vegetation and exclude exotic species. Native

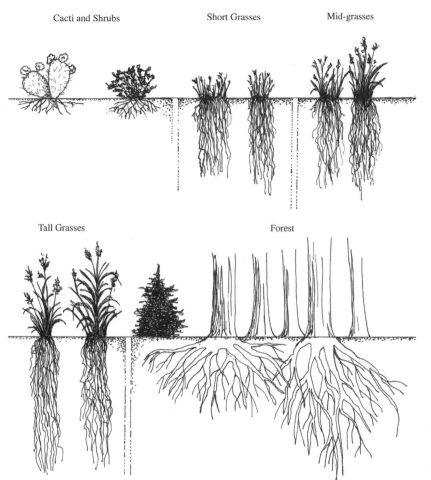

Cacti and Shrubs   Short Grasses   Mid-grasses

Tall Grasses   Forest

Figure 5-4 • The North American prairie can be divided into three broad categories of grasses (short grasses, mid-grasses, and tall grasses) that grow between a desert biome area in the west to a forested biome in the east.

American plains tribes traditionally set fires to promote vegetative growth or to drive wildlife for hunting. Fire-suppression efforts since the early 1900s have changed the species composition of many fire-adapted grassland communities and have reduced prairie biodiversity. The conversion of tallgrass prairie to agricultural lands had left only an estimated 1 percent of the prairie undisturbed by settlers.

PAMPAS

The pampas is a vast temperate grassland environment located in southeastern South America. The word "pampa" is a Quechua Indian term meaning "flat surface." The vast grass-covered plains that make up the South American pampas extend westward across central Argentina from the Atlantic coastline to the foothills of the

Tallgrass Prairie National Preserve

Interest in conservation and restoration of prairie as a heritage ecosystem began a few decades later in the United States. The Tallgrass Prairie National Preserve, in the Flint Hills region of Kansas, is the country's only national park devoted to the preservation of the tallgrass prairie. The preserve was signed into existence in 1996 as part of the Omnibus Parks and Public Lands Act. Once private land, the preserve's 4,437 hectares (11,500 acres) were purchased by the nonprofit National Park Trust, which will manage the land in partnership with the National Park Service. Other ongoing prairie-restoration projects are Konza Prairie (Kansas) and Walnut Creek (Iowa).

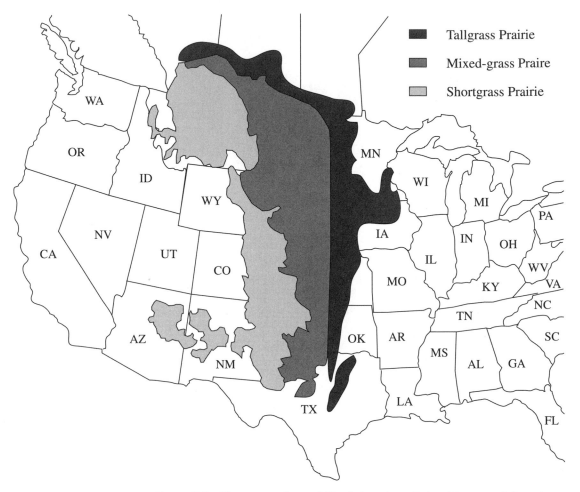

Figure 5-5 • The grasses of central North America, called prairies, extend through much of the midwestern part of the United States and south-central Canada.

Andes mountains. The Argentine pampas, which cover an area of approximately 760,000 square kilometers (295,000 square miles), can be divided into two separate sections. The western dry section, with sandy deserts and dry, scrubby vegetation, is largely uninhabited by people. Rainfall is much higher in the eastern pampas where vegetation is more abundant than in the west.

Grass plants are the dominant form of life in the pampas just as they are in all grassland ecosystems. Grasses are perfectly suited to low precipitation and high temperatures. Their root systems form dense mats that can survive droughts and fire and hold soil together. Few trees can survive in this kind of environment because of drought, fire, and the high winds that often roar across the open plains. Characteristic animals of the pampas include hawks, a number of burrowing rodent species, and guanacos, large herbivorous members of the camel family.

Elephant

Zebra

Wildebeast

Cheetah

Kangaroo

Rhea

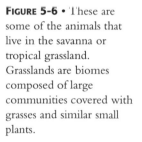

FIGURE 5-6 • These are some of the animals that live in the savanna or tropical grassland. Grasslands are biomes composed of large communities covered with grasses and similar small plants.

STEPPE

Another type of temperate grassland is the steppe. It too is dominated by grasses with few if any trees. The vast steppe lands stretch from southern Russia, through Kazakhstan and southern Siberia, west to Mongolia, and south to include western China. The steppe climate is characterized by light rainfall in the spring and early summer for the growing season. It has a dry, sunny, late summer which supports the ripening and harvesting of crops. Wild camels live on the Mongolian steppe.

Tropical Grasslands or Savannas

Tropical grasslands, called savannas, are located near the equator in central and southern Africa and in parts of South America. Other smaller savannas are located in Australia and China.

The savanna is a tropical or subtropical habitat that consists mainly of grasses, bushes, and scattered small trees. The word savanna comes from the Spanish word "zavanna," which means a "treeless plain." Savannas are often located between equatorial rain forests and desert regions. The year-round temperatures are high, ranging from 38° to 45°C (70° to 85°F), and the precipitation, most of it falling only once or twice a year, averages about 12 centimeters (30 inches) in a year.

For the most part, savannas are located primarily in Africa and cover large landscapes of that continent. They include the Sahel and Sudan savannas. The Sahel in the north flanks the Sahara Desert. The Sudan savanna is in the south where the trees are larger than those in the north. Some of the most popular wildlife refuges in the world include Tanzania's Serengeti Plains National Park and Kenya's Tsavo National Park, both well-known African savannas for *ecotourism*.

The grasses and shrubs play an important role in the ecosystem of the savanna. Short and tall grasses, underground stems and roots, and other vegetation support large groups of grazing herbivores, including wildebeest, gazelles, and zebras in Africa. During the year, many of these African animals migrate in herds numbering in the thousands in search for food and water. In turn, the herbivores provide food for predators such as cheetahs, lions, leopards, hyenas, and wild dogs, as well as *scavenger* birds such as vultures. In Australia, the grazers include rabbits, kangaroos, and wallabies. Other herbivores include giraffes, antelopes, and elephants. The savanna also supports a variety of birds, lizards, and insects. The savanna is home to the ostrich, the largest bird,

FIGURE 5-7 • The Serengeti National Park is located in a vast subtropical grassland in Tanzania near the Kenya border.

Serengeti Plains National Park

The largest and oldest of Tanzania's 12 national parks, this is an area where *poaching* can be a problem, particularly for elephants and the black rhinoceros. However, every effort is being made by the Tanzanian government to protect these animals and conserve this unique park.

The Serengeti National Park, located in the vast subtropical grassland or savanna in east Africa, is one of the most famous wildlife parks there. Its name is derived from a Maasai word, "siringet," meaning "wide open space," or "extended place." Established in 1951, the park is about 14,000 square kilometers (5,400 square miles) or about the size of the state of Connecticut. It is located near Lake Victoria, one of the largest lakes in the world. The Serengeti's climate is usually warm and dry. The main rainy season is from March to May, with short rainy seasons from October to November. The yearly rainfall can average between 500 millimeters (20 inches) to 1,200 millimeters (48 inches).

The Serengeti Research Institute, founded in 1962, has provided valuable data for the management and conservation of this park. Probably more is now known about the dynamics of the Serengeti ecosystem than any other ecosystem in the world.

Serengeti Park has a unique combination of diverse habitats which support more than 30 species of large herbivores and nearly 500 species of birds. The Serengeti savanna is home to more than two million wildebeest, half a million Thomson's gazelles, and a quarter of a

The African elephant is the largest land animal on Earth and an endangered species.

million zebras. The park has the greatest concentration of plains game in Africa. Hundreds of thousands of wildebeest and zebra provide a unique spectacle for tourists as the animals migrate from the southeastern plateaus of the Serengeti northward in search of food and water. Apart from the rhinos, which have been decimated by poachers, the wildebeest and buffalo populations have multiplied, benefiting the main predators—lions, cheetahs, and hyenas. Other animals include giraffes, warthogs, jackals, ostriches, lizards, oryx, eagle leopards, hippopotamuses, and vultures.

which can weigh more than 160 kilograms (350 pounds) and reach a height of 2 meters (6 feet).

Environmental Concerns of Grasslands

Today, many grasslands are suffering from wind and water erosion of topsoil, the overuse of pesticides, the depletion of *aquifers*, poaching, and the invasion of nonnative or exotic species of plants. However, the

Termites

In the open grasslands there are termite mounds that contain thousands of termites. The mounds, which are quite hard, are built of sand particles held together by the saliva and excrement of the termites. The termites feed on dead vegetation that is broken down by fungi that also grow in special chambers of the mound. Fresh air is drawn into the mound through many tunnels in the nest. The tunnels allow hot air and carbon dioxide to be vented out of the nest. The constant circulation of air maintains a constant temperature in the mound which helps the queen termite to lay her eggs.

major threats to the pampas and other grassland areas are farming and overgrazing by cattle, horses, and sheep. The grain crops, such as alfalfa and wheat, that have replaced the native grasses cannot hold soil in place very well because their root systems are very shallow. This results in soil erosion, which reduces the overall fertility of soils. In addition, overgrazed grasses which are constantly chewed down often do not regenerate, making the bare soil even more prone to further erosion. Overuse of the land for farming and grazing has caused desertification in many parts of Africa. This condition was the cause of the infamous 1930s Dust Bowl in the United States. The increase in human population in the savanna has also led to a loss of wildlife habitat forcing more wildlife to live in smaller and smaller areas.

DESERTS

Deserts are the driest biomes on Earth because of the lack of water. Typical deserts occur between 15° and 35° north and south latitudes. Precipitation in the world's deserts is very low, averaging less than 25 centimeters (10 inches) per year.

Termite hills or mounds are found in open grasslands and in the Litchfield National Park in Australia. Termite nests in Australia exude streams of carbon dioxide into the atmosphere. Termites account for a significant amount of all carbon dioxide released on land through decomposition. (Courtesy of Degree Confluence Projet/ Jean-Michel Weil)

The principal global deserts include the Sahara and Libyan in north Africa, the Kalahari in southwest Africa, the Arabian in the Middle East, the Gobi in central Asia, the Patagonia in South America, and several deserts in central Australia. Four desert regions make up the North American Desert: the Great Basin, the Mojave, the Sonoran, and the Chihuahuan.

The Sahara Desert is the largest desert in the world. It covers more than 9 million square kilometers (3.5 million square miles). The driest desert is the Atacama in South America. Its average rainfall is less than one-half a centimeter (or about 0.4 inches) a year. By contrast, many tropical rain forests average about 203 centimeters (80 inches) a year. Deserts are not all sand; for example, only 10 percent of the Sahara is sand. The rest of the landscape is rocky.

Tropical or hot deserts are located at subtropical levels and low elevations. Extremes in temperature often occur in deserts, ranging from very hot daytime conditions to near-freezing temperatures at night. In tropical deserts, such as the Sahara Desert in northern Africa, temperatures are extremely hot all year round. Other tropical deserts include the Sonoran in the United States, Chihuahuan in Mexico, and the Kalahari in Africa.

Temperate latitude or cold deserts are found too far north or too high in elevation to be consistently hot. As an example, in the Mojave Desert in the United States, the temperatures are hot during the summer but relatively cold in the winter. Winter temperatures can reach as low as 0°C (32°F). Other temperate deserts include the Patagonia in South America and the Gobi in China. Both tropical and temperate deserts have something in common—they are inhospitable to most forms of life.

Life in the Desert

Deserts constitute the least productive type of ecosystem because the limited and irregular water supply supports only small populations. Desert organisms are therefore adapted for surviving in these extremely dry and sometimes very hot conditions. Some animals, such as the kangaroo rat, survive by ingesting moisture from the seeds they eat without drinking water. Other animals are principally *nocturnal*. They have the ability to be active during the night when it is cooler. Spadefoot toads burrow into mud banks and emerge from their holes only during periods of rainfall or cooler conditions. Other animals inhabiting the desert include lizards, snakes, tortoises, gerbils, camels, tarantulas, scorpions, and a variety of insects. Harris's hawk and the red-tailed hawk are the principal predators in North American deserts. Other predators include roadrunners, kit foxes, bobcats, gopher snakes, and owls. Poisonous desert animals include rattlesnakes, scorpions, gila monsters, and coral snakes.

Many desert plants have exterior surfaces that enable them to reduce moisture loss to the dry desert air and remain dormant for long

DID YOU KNOW?

During the time of the Roman Empire, the present Sahara Desert was actually a fertile area consisting of grasslands and some trees. Cereal crops were farmed for the Romans and the local inhabitants.

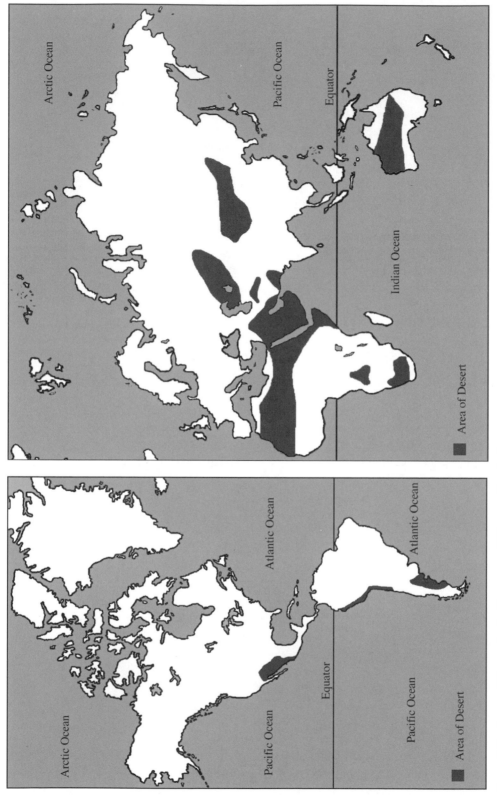

FIGURE 5-8 • Areas of Desert in the Eastern and Western Hemispheres

OREGON

IDAHO

WYOMING

CALIFORNIA

Great Basin
Desert

COLORADO

Painted
Desert
Community

Mohave
Desert

ARIZONA

Colorado
Desert

Arizona
Sonoran
Desert

NEW MEXICO

TEXAS

Chihuahuan
Desert

FIGURE 5-9 • North American Deserts

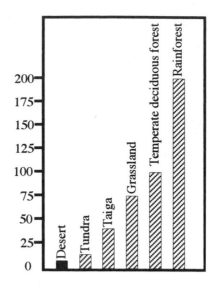

FIGURE 5-10 • Average Annual Rainfall of Biomes (cm/year)

dry spells. Cactuses are examples of well-adapted desert plants because they have small or needle-like leaves that help the plant conserve water. The prickly pear cactus has shallow, wide-spreading roots that can soak up moisture even from just a light rain that dampens the surface. The roots of the creosote bush produces toxins which kill off competitive plants that have grown too close. The roots of the mesquite cactus can extend more than 20 meters (60 feet) to tap water. Some varieties of desert plants have developed the ability to flower rapidly and reproduce during brief periods of moisture.

Historically, many desert cultures of people have existed throughout the world. Indigenous or aboriginal groups (Australian Aborigines) occupied vast arid portions of the Australian continent for thousands of years. In the southwestern United States, many generations of the native American Pueblo peoples successfully inhabited desert regions. The Near East was well known for the nomadic tribes that roamed the Arabian and Persian deserts. The Kalahari Desert of South Africa is the home of the Bushmen cultures.

Environmental Concerns of Deserts

Overcultivation and overgrazing of domestic animals have damaged once fertile areas in the desert. Of the 2,000 known cacti species, approximately 20 are at risk of extinction, primarily because of extensive collecting by humans and damage caused by grazing animals. Strong windstorms have carried away fertile soil. Drought conditions in some African countries have caused starvation for many people who live on the land. *Salinization* has occurred in many irrigated farm areas in central Asia and in the San Joaquin Valley of California.

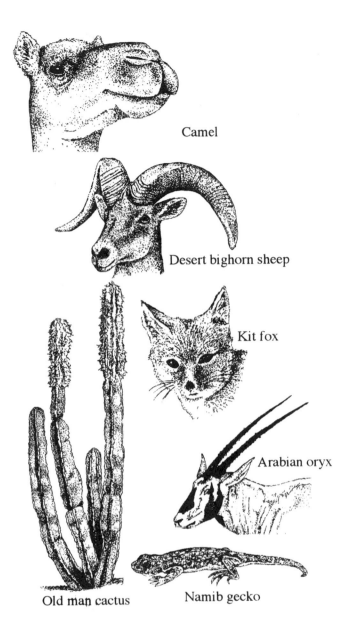

Camel

Desert bighorn sheep

Kit fox

Arabian oryx

Namib gecko

Old man cactus

FIGURE 5-11 • These are some of the animals and plants that live in a Desert Biome.

CHAPARRAL

A biome sometimes referred to as a scrubland is known as the chaparral. The chaparral is dominated by scrub oaks, evergreen bushes, and thorny shrubs. The climate consists of long, hot, dry summers with low precipitation and cool, rainy winters. Chaparrals are found in central and southern California and throughout countries in southern Europe along the coast of the Mediterranean Sea. Other chaparrals are found in Africa, Chile, Mexico, and parts of Australia. The chaparral is home to mostly small animals, such as lizards, snakes, rabbits, and birds, and some browsing animals, such as sheep and goats.

Cacti

Cacti (*Cactaceae*) are generally adapted to life in desert or near-desert environments. There are about 2,000 species of cacti, most of which are native to the southwestern United States, Mexico, Central America, and the southernmost countries of South America. However, of the 2,000 cactus species known, more than half grow in Mexico and southwestern United States. Cacti are not limited to the desert; they also grow in savannas, along seashores, on islands, and in the rain forest.

Cacti have several adaptations to help them conserve water and thrive in arid regions. Most cacti are succulents—plants with swollen, fleshy stems that store large amounts of water. The stems also serve as the sites where photosynthesis occurs. In most plants, photosynthesis and *transpiration* occur in leaves, structures that are absent in cacti. Instead of leaves, many cacti have spines which help conserve water and protect cacti from predators. The root structure of cacti also differs from that of many other plants. Instead of penetrating deep into the soil, as do the roots of many trees, the roots of cacti spread out laterally, near the soil's surface. By remaining near the surface, the roots can quickly absorb water made available through infrequent rains.

Cacti vary greatly in size and shape. The largest, the giant saguaro (*Cereus giganteus*), is native to the United States and receives protection within the boundaries of the Saguaro National Monument in Arizona. Some cacti, such as the night-blooming cereus and the endangered Chisos Mountain hedgehog cactus, produce beautiful flowers. Many cacti, such as the night-blooming cereus and the peyote, contain chemicals that are used to make medications or beverages having alcohol-like qualities. Others, such as the prickly pear, produce edible fruits. These diverse characteristics have made many species of cacti attractive to collectors of rare and exotic plants.

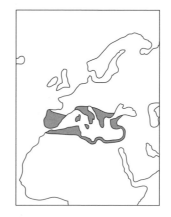

FIGURE 5-12 • The coastline chaparral of the Mediterranean is dominated by scrub oaks, evergreen bushes, and thorny shrubs. Destruction of the chaparral has resulted from clearing the land for agriculture.

Fast-burning fires can occur during the dry, hot summer season and sweep through large areas of the chaparral. However, the fires are necessary for some plants to regenerate from roots, bulbs, and seeds lying deep beneath the surface.

Environmental Concerns of the Chaparral

Destruction of the chaparral has resulted from clearing of the land for agriculture and the overuse of pesticides and fertilizers which can damage sensitive ecosystems. Clearing of chaparral vegetation on slopes allows runoff thereby producing hillside erosion and flooding.

TUNDRA

After deserts, the tundra is the driest biome on Earth. The tundra lies north of the taiga forests and is found worldwide within the Arctic Circle and the Antarctica. It is a vast, treeless biome characterized by a very cold, very dry climate and vegetation such as lichens, mosses, herbs, grasses, sedges, rushes, and small shrubs. An alpine tundra is located on high mountains.

The tundra environments have the lowest average annual temperatures of all biomes because of the Earth's curvature. The sunlight shining at higher latitudes, such as in the tundra, is spread over a wider area than the sunlight shining near the equator. The higher the latitude, that is, the further away from the equator and closer to the North and South poles, the colder is the climate.

The climate in the tundra is characterized by low temperatures, from −25°C (−13°F) in the winter, to about 6°C (43°F) in the summer, and very little precipitation—about 12 centimeters (4:7 inches) per year, mostly in the form of snow.

Winters are very long in the tundra, and there are many months of twilight and darkness when the sun is below the horizon for the entire day. As a result, the growing season is very short. During the brief, cool summer season, plants sprout and flower quickly. Only a few plants and animals, which possess structural and behavioral adaptations that help them survive the hostile climate, can live in such inhospitable conditions.

Permafrost

Due to the lack of vegetation in the tundra, the soil quality is poor. Below the surface of the ground is a layer of permanently frozen soil called permafrost. An estimated 20 to 26 percent of Earth's land area is covered by permafrost which probably formed during the last *ice age*. Depending upon the temperature range in an area, permafrost may be discontinuous or continuous. Permafrost can range in thickness from less than 1 meter (3.3 feet) to more than 400 meters (1,312 feet). Virtually all of Antarctica and the country of Greenland rest atop continuous permafrost. Much of Canada, Alaska, Siberia, and the portions of northern Europe and Asia located above 45° to 50° latitude also have large regions of continuous permafrost. Portions of North American, European, and Asian countries located at (or just below) 45° to 50° latitude generally have discontinuous permafrost—regions of permafrost intermingled with regions of thawed ground or bogs.

Tundra Plants

In many places, the top soil layer above the permafrost thaws a little during the summer. The thickness of this "active layer" depends on the average temperature of the region, the slope of the land, and the soil's

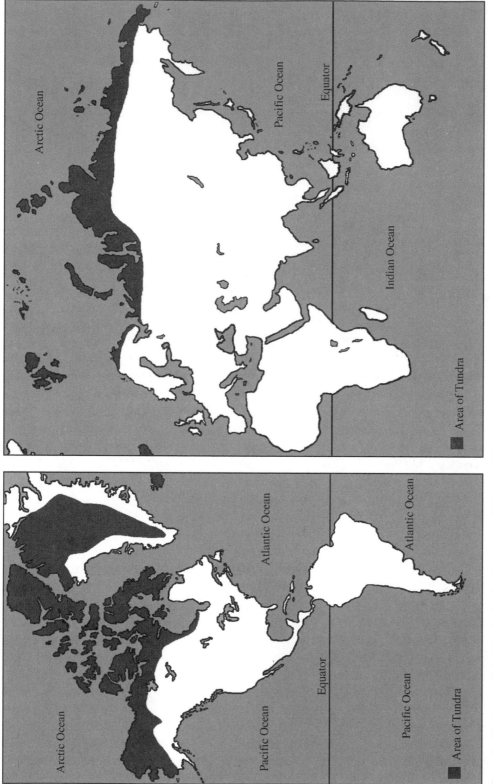

Figure 5-13 • Areas of the Tundra in the Eastern and Western Hemispheres

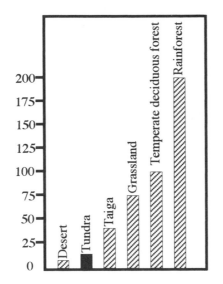

Figure 5-14 • Average Annual Rainfall of Biomes (cm/year)

exposure to sunlight. It is in this thawed soil that plants can grow, sometimes in large enough numbers to warm the soil and the permafrost below it. However, the permafrost and the low precipitation prevent the rooting and growth of forests in the tundra. The plants that do grow in the tundra—lichens, mosses, shrubs, sedges, and dwarf trees—adapt by growing very close to the ground, protected from the battering winds that whip across the open territory. The presence or absence of plants, which along with lichens and ice algae are tundra producers, plays a vital role in determining which animal species can survive in a permafrost region.

Tundra animals adapt to the difficult conditions as well. Many of the year-round animal residents, including reindeer, caribou, wolves, polar bears, and musk oxen, have thick skin and fur to protect them from the cold. There are marine species as well, such as walruses and seals. Others, such as snowshoe hares, foxes, and snowy owls, are adorned with white fur or feathers to help them blend in with their surroundings and avoid predators. Still other animals, such as mice, voles, lemmings, and other small rodents, build burrows under the snow. By far the most abundant animals in the tundra are insects, especially in the southernmost regions where boggy soils, streams, rivers, and lakes are common. Huge swarms of mosquitoes and black flies make life almost unbearable for other animals. The tundra is also home to a great variety of birds. Some, such as the snowy owl, live there all year. Others, including many species of ducks and geese, nest in the tundra during the summer and migrate south by the millions during the harsh winter months when temperatures plummet and food sources have disappeared. Human populations in the Arctic include the Inuit peoples and others.

Mammoths

In 1997 a complete specimen of a woolly mammoth was discovered in the permafrost of Siberia. All of the body parts were intact. The mammoth roamed the Siberian grasslands in large herds about 20,000 years ago.

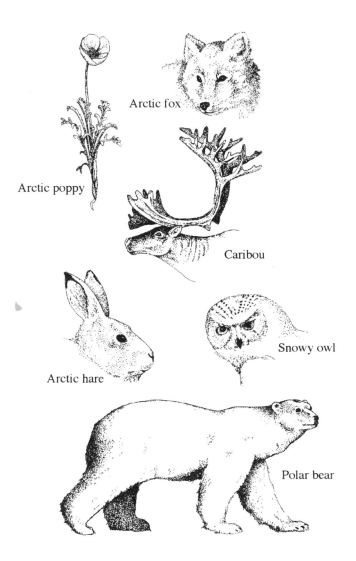

FIGURE 5-15 • These are some of the animals and plants that live in a Tundra Biome.

Environmental Concerns of the Tundra

Studies indicate that temperatures in parts of the Alaskan, Canadian, and Siberian tundra are increasing by about 1°C (1.8°F) every decade, possibly as a result of global warming. If this trend continues, the top 10 to 15 meters (33 to 50 feet) of the thicker permafrost regions could thaw and significantly change the tundra ecosystem. Thawing of permafrost also leads to subsidence, which destroys roads, homes, and a fragile ecosystem that can easily be disrupted. Moreover, as permafrost thaws, trapped methane gas is released into the atmosphere. Since methane is a greenhouse gas, its release could contribute to greater global warming. Many scientists support abandonment of the search for additional petroleum deposits in tundra regions because the tundra is such a fragile ecosystem. They also discourage construction of additional facilities and pipelines for collecting and transporting oil in these areas.

Vocabulary

Aquifer Porous rock or soil reservoir through which water passes and can be drawn up through wells.

Climax The final stable state in the development of an ecosystem.

Ecotourism Tourism of interesting and unusual ecological areas.

Ice age A long period of time when Earth temperatures were cool and large areas of the surface became covered with ice.

Nocturnal An organism that is active during the night.

Poaching Illegal hunting of wildlife.

Salinization A process in which soil or water becomes more salty. Evaporation of water leaves salt as a crust on dry surface.

Scavenger An organism that feeds on dead matter.

Transpiration Loss of water from a plant through its stomata, especially in the leaves.

Activities for Students

1. Conduct research on the biomes where the bulk of human evolution took place. How do our physical adaptations reflect these biomes? What type of provisions have we made in order to survive in other habitats?

2. Pretend that you are planning an ecotourism adventure in the savannas of eastern Africa. Learn more about Serengeti National Park and create an itinerary for a safari. What would you want to see and do?

3. Proposals have been made in the government that we should begin exploration for oil in the Arctic Wildlife Refuge in Alaska. Consider the issue and write a letter to your local congressperson sharing your opinion on this possibility.

Books and Other Reading Materials

Brown, Lauren. *Grasslands*. New York: Alfred A. Knopf, 1985.

Dybas, Cheryl Lyn. "Houses Bow to Thawing Permafrost," National Science Foundation, 1995 cited September 3, 1997, http://www.nsf.gov/stratare/egch/nws195.htm.

Foster, Lynne. *Adventuring in the California Desert*. Sierra Club Adventure Travel Guide. San Francisco: Sierra Club Books, 1997.

Lazaroff, David Wentworth. *Arizona-Sonora Desert Museum Book of Answers*. Tucson, Ariz.: Desert Museum Press, 1998.

Man, John. *Gobi: Tracking the Desert*. New Haven, Conn.: Yale University Press, 1999.

Manning, Richard. *Grassland: The History, Biology, Politics and Promise of the American Prairie*. New York: Penguin USA, 1997.

Savage, Stephen. *Animals of the Desert*. Austin, Texas: Raintree/Steck Vaughn, 1997.

Tundra Keller, Edward A. *Environmental Geology*. New York: MacMillan, 1995.

Van Dyne, G. M. *Grasslands, Systems Analysis and Man*. Cambridge, England: Cambridge University Press, 1980.

Zwinger, Ann Haymond. *The Mysterious Lands: A Naturalist Explores the Four Great Deserts of the Southwest*. Phoenix: University of Arizona Press, 1996.

Websites

Desert Research Institute, http://www.driedu

Mojave National Preserve, http://www.nps.gov/moja/mojadena.htm

Postcards from the Prairie, http://www.nrwrc.usgs.gov/postcards/postcards.htm

Steppe information, http://www.worldwildlife.org/ action/factsheetslite/mongolia/bio.htm or http://www.lternet.edu/network/sites/ 05_cpr.html

Tallgrass Prairie National Preserve, http://www. nps.gov/tapr/home.htm

University of California, Berkeley, World Biomes: Grasslands, http://www.ucmp.berkeley.edu/glossary/ gloss5/biome/grasslan.html

Worldwide Fund for Nature: Grasslands and Its Animals, http://www.panda.org/kids/wildlife/ idxgrsmn.htm

CHAPTER 6

Water Biomes

Water biomes are part of Earth's hydrosphere and they are known as aquatic biomes. Freshwater biomes include streams, rivers, ponds, lakes, and marshes. The water in these biomes contains little dissolved salt. Ocean, seas, estuaries, and saltwater marshes are known as *marine* biomes. They contain higher concentrations of dissolved salt than freshwater. Water biomes exist in close connection with land biomes. The boundaries between the two constitute extremely important ecological zones.

Refer to Chapter 2 for more information about the hydrosphere.

PLANKTON

Plankton are the basic link in freshwater and ocean biome food webs. Plankton are tiny marine and freshwater organisms that drift at or near the water's surface. Some plankton do have structures for locomotion; however, most are dependent upon wind, tides, and water currents for their movement. Plankton are most plentiful in the *littoral* zones of ponds, lakes, and oceans which are areas along the shores. In the littoral zones, sunlight reaches the bottom of lakes, ponds, and the ocean floor and provides energy for plankton photosynthesis. Water near the shore appears green because of the presence of green plankton. Water in the deeper areas of lakes, ponds, and oceans appears blue because of the low number of plankton.

Refer to Chapter 3 for information about food webs.

Plankton are divided into two major groups: phytoplankton and zooplankton. Phytoplankton that live in the *pelagic zone* include

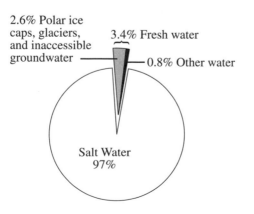

2.6% Polar ice caps, glaciers, and inaccessible groundwater

3.4% Fresh water

0.8% Other water

Salt Water 97%

FIGURE 6-1 • Freshwater reserves represents between 3 and 3.5 percent of all Earth's water. However, much of it is stored in ice caps and glaciers making these freshwater resources unavailable for human needs.

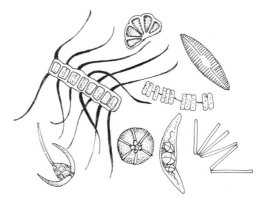

FIGURE 6-2 • Plankton are divided into two major groups: phytoplankton (autotrophic) and zooplankton (heterotrophic).

microscopic organisms that derive their nutrition and energy from photosynthesis; for example, some species of bacteria, fungi, and algae. Phytoplankton are producers and thus form the base of aquatic food chains. Zooplankton are consumers. They consist primarily of protozoa, small crustaceans such as krill, worms, jellyfishes, and the eggs and larvae of many aquatic animals. Most zooplankton drift along and feed on phytoplankton; however, some may feed on other zooplankton. These small organisms, in turn, are used as a food source by many larger fishes and aquatic mammals.

Refer to Chapter 3 for more information about producers and consumers.

Plankton are extremely numerous. In fact, a single liter of pond or lake water may contain more than 500 million planktonic organisms. When marine plankton exist in such great numbers, they can actually cause an apparent change in the color of the water. Red tides, for example, result when billions of algae known as dinoflagellates (a type of phytoplankton) are present in ocean water. These tides can be dangerous because dinoflagellates produce a toxin that is poisonous to fish, shellfish, and humans.

FRESHWATER BIOMES

Rivers

Freshwater has its origin in precipitation such as rain and snow. Some of the fallen rain or snow is soaked up by the ground, but much of the water runs off and flows into streams and other tributaries. A stream typically begins high in the mountains. The headwater or beginning of a stream is usually very steep and narrow. The water in the stream is very cool and crystal clear, and it flows very fast. Very few organisms can live in a fast moving stream. However, some living organisms can resist the pull of the current; some plants can grow strong roots that anchor the plants to the bottom of the stream. Other living organisms, such as insect larvae, small invertebrates, and worms, can burrow under rocks and stones. Freshwater shrimp can cling to the rocks with their claws. As the stream slows down and widens, the current decreases.

Refer to Chapter 1 for more information about freshwater areas such as rivers, ponds, bogs, wetlands, and marshes.

TABLE 6-1	Major Rivers of the World	
Name	Location	Approximate Length in Kilometers (miles)
Nile	Africa	6,693 (4,160)
Amazon	South America	6,436 (4,000)
Chand Jiang (Yangtze)	Asia	6,378 (3,964)
Mississippi/Missouri	North America	6,017 (3,740)
Huang (Yellow)	Asia	5,463 (3,395)
Ob-Irtysh	Asia	5,409 (3,362)
Amur	Asia	4,415 (2,744)
Lena	Asia	4,399 (2,734)
Congo	Africa	4,373 (2,718)
Mackenzie	North America	4,240 (2,635)

The gentler waters, rich in dissolved oxygen and abundant nutrients, can support a variety of fish, crustaceans, and amphibians. Mammals such as otters, hippopotamuses, water snakes, and crocodiles live along the streams and riverbanks. Plants found along riverbanks include mosses, ferns, and liverworts.

Ponds and Lakes

Ponds and lakes are bodies of still water lying in depressions. Precipitation and groundwater collect and accumulate in the depressions forming ponds or lakes. Unlike rivers or streams, water is still or moves very slowly in ponds and lakes.

PONDS

Ponds are usually smaller than lakes and may not exist year-round. As an example, some ponds dry up in midsummer as the water evaporates from the summer heat. The quiet waters of ponds support a variety of living organisms. The shoreline includes grasses, trees, and other plants which provide habitats for other animals. Many birds such as redwing blackbirds, ducks, and geese nest along the shore. Frogs lay their eggs in the shallow areas near the shore. Pads of water lilies live on pond bottoms. Bottom dwellers such as snails and crayfish feed on the nutrients that drift down from the surface. Insects include dragonflies, beetles, and water striders. Plants grow throughout the water. Microscopic algae (phytoplankton), floating in the water, make food through the process of photosynthesis. Consumers such as fish, turtles, frogs, and animals use the nutrients and oxygen provided by the plants and other producers.

The Hoover dam is one of the largest dams in the world. It spans the Colorado River and provides water and electricity for Arizona, Colorado, and Nevada. (Courtesy of U.S. Department of Interior. Bureau of Reclamation, Andrew Pernick, Photographer)

LAKES

Lakes form in many ways. Some lakes were created by tectonic activity, or movement in Earth's crust, such as earthquakes and volcanoes. After a volcanic eruption, water can collect in the empty craters of the volcanoes forming lakes. People have also created lakes, known as reservoirs, to store water. One of the largest human-made reservoirs in the United States is Lake Mead in Nevada. It lies behind Hoover Dam on the Colorado River. Retreating glaciers also dug out depressions in the ground which became lakes. As an example, the Finger Lakes in New York state were formed this way.

Like a pond, a lake supports a large diversity of species. In the littoral zone, the living organisms are similar to those of a pond environment. Onshore, there are nesting birds such as loons, kingfishers, ducks, geese, and other wading birds. There are populations of frogs and turtles. Reeds and cattails are rooted in the muddy waters. Pond lilies float at the surface. The lake is home to many kinds of insects such as water beetles, dragonflies, and caddis flies. Fish include catfish, pike, bass, and sturgeon.

Unlike ponds, solar energy does not reach the bottom of the middle of a deep lake. There is too little light for the process of photosynthesis by microscopic algae and other producers. As a result, fewer organisms live in the deep water. This area is known as the *benthic zone*. Dead organisms that drop to the bottom are decomposed by bacteria. Other bottom dwellers include freshwater clams and insect larvae.

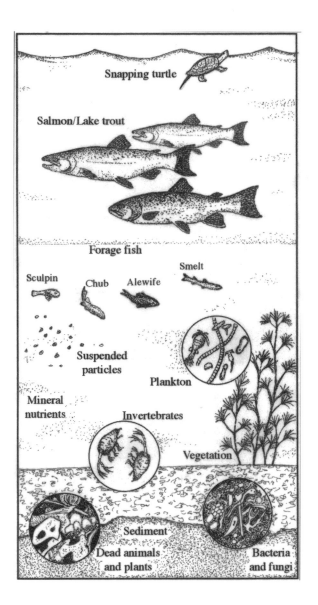

FIGURE 6-3 • This is an illustration of a simplified aquatic ecosystem showing producers and consumers. (Courtesy of Michael Kamrin, Michigan State University, *Ecotoxicology for the Citizen*)

ENVIRONMENTAL CONCERNS OF RIVERS, PONDS, AND LAKES

Along riverbeds, the impact of land erosion can disturb aquatic habitats. Toxic chemicals and other pollutants can be dumped into the water and harm or kill wildlife. Excessive nutrients in ponds and lakes can cause unwanted growth and can lead to the *eutrophication* of lakes and ponds.

Freshwater Swamps, Marshes, and Bogs

A habitat that is saturated with water all or part of the year is known as a wetland; however, a few wetlands are dry most of the year. Wetlands include bogs, coastal saltwater marshes, freshwater marshes, swamps,

Lake Baikal

The world's oldest and largest freshwater lake, located in south-central Siberia, Lake Baikal contains about 20 percent of Earth's liquid freshwater, and almost 80 percent of the freshwater within the former Soviet Union. Located near the Mongolian border, Lake Baikal measures almost 640 kilometers (400 miles) in length, averaging 85 kilometers (50 miles) in width, and reaches depths as great as 1,740 meters (5,712 feet). Lake Baikal is supplied with freshwater by more than 300 rivers and streams. The Serenga River is one of the major tributaries that empties into the lake. It has been estimated that almost 2,500 plant and animal species live in or around the lake. Of these, more than 1,000 species are found nowhere else on Earth. One of the animals found only in the Lake Baikal region is the Baikal or Nerpa Seal. It is the only known species of freshwater seal.

Other large mammals include elks, brown bears, moose, and deer. Because of pollution, Lake Baikal has become an environmental concern.

FIGURE 6-4 • Lake Baikal is the oldest, largest, and deepest freshwater lake on Earth. It contains about 20 percent of Earth's liquid, fresh water and almost 80 percent of the former Soviet Union's freshwater supply. However, pollution has begun to slowly take its toll on plant and animal species in the region.

vernal pools, and fens. Each wetland is characterized by a variety of animal and plant communities and varies in types of soil, topography, climate, and water content. Wetlands are found on most continents, even in the tundra. In the United States, wetlands are found in every state, from the coastal marshes of Alaska to the mangrove forests of Florida.

BOGS

Bogs, also known as peatlands, are simply wetlands that have organic soils consisting of peat—the partially decomposed remains of plants and animals. Bogs depend primarily on precipitation as their water source and are usually acidic and rich in plant material. Grasses, sedges, and mosses constitute typical bog vegetation. Insect-eating pitcher plants, sundew, and Venus's-flytrap live in the water-logged acidic soil.

A bog can exist for a long period of time; however, some disappear as they fill with sediment, become a habitat for shrubs and trees, and eventually develop into a forest ecosystem. Bogs are found in colder regions of the world where the temperatures and limited oxygen supply in the water discourage the breakdown of organic material. A fen is another type of wetland that is less acidic than bogs because the water contains calcium and magnesium. Peat also accumulates in a fen.

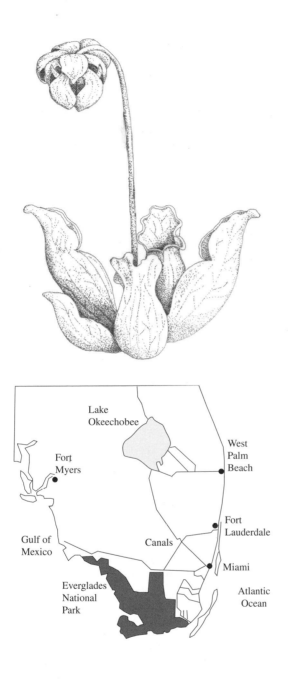

FIGURE 6-5 • One kind of pitcherplant grows in bog-like soils. Bogs are wetlands that receive their water only from precipitation—not from streams. Other bog plants include the Venus's flytrap and sundew.

FIGURE 6-6 • The diverse ecosystems of the Everglades provide habitat for abundant wildlife, including rare and colorful birds, manatees, turtles, snakes, alligators, the Florida panther, and crocodiles.

FRESHWATER MARSHES

Freshwater marshes are familiar to most Americans. They make up nearly 90 percent of the wetlands in the United States. The Florida Everglades is the largest freshwater marsh in the United States. Most of the water that flows into the Everglades comes from Lake Okeechobee which lies north of the Everglades.

Marshes are open areas, usually with few trees and shrubs, which provide food supplies for migratory birds and other wildlife. Sedges, reeds, rushes, and cattails are common in freshwater marshes. The water in a marsh fluctuates, rising during the rainy season and disappearing during dry periods. Remaining freshwater wetlands in suburban and urban areas are valuable for purifying storm water and reducing flooding.

Prairie potholes are one common type of wetland. Prairie potholes are saucer-shaped depressions formed by retreating glaciers during the ice age. Although inundated with water for only a short period of time each spring, they play a vital role in aquatic and wildlife habitat. Prairie potholes, located in the Upper Plains states, are often called the "duck factories" of America because of their importance to the livelihood of ducks and other migratory birds. In addition to supporting waterfowl and other birds, prairie potholes also absorb rain, the melting of snow, and floodwaters and release the water slowly throughout the watershed, thereby reducing the risk and severity of downstream flooding.

SWAMPS

A swamp is a wetland in which the dominant plant life is woody vegetation such as trees or shrubs; specific swamps are cypress swamps and mangrove swamps. As a wetland, a swamp remains at least partly covered by water throughout the year and provides habitat to a variety of organisms, including waterfowl, migratory birds, amphibians, and often large insect populations. Unlike bogs, swamps generally lack peat deposits on their floors. Depending upon their water source, swamps may be freshwater or marine; marine swamps are tidal or nontidal in nature. Swamps are shrubby or forested wetlands, located in poorly drained areas on the edges of lakes and streams. Forest swamps primarily exist in the *floodplains* of major river systems. Swamps that existed 250 million years ago developed into present-day coal reserves in many locations of the United States.

Vernal pools are small, isolated wetlands which retain water on a seasonal basis. The pools are vital to the survival of amphibians; nearly 50 percent of the amphibians in the United States breed primarily in vernal pools because the pools are too shallow to support fish, the major predator to amphibian larvae. Vernal pools are also home to many endangered and rare plant species. The unique environment of vernal pools provides a habitat for numerous rare plants and animals which are able to survive these harsh conditions. Many of these plants and animals spend the dry season as seeds, eggs, or cysts and then grow and reproduce when the ponds are flooded again. In addition, birds such as egrets, ducks, and hawks use vernal pools as a seasonal source of food and water.

ENVIRONMENTAL CONCERNS OF WETLANDS

Over the past 200 years, human activities have directly caused the destruction of valuable wetlands by draining, filling, and flooding. By 1985 between 55 and 65 percent of the available wetlands in Europe and North America had been drained. In the United States, as an example, the U.S. Environmental Protection Agency (EPA) reported that about 50 percent of the wetlands in the lower 48 states were lost between the late 1700s and the mid–1980s.

MARINE BIOMES

The world's oceans cover about 361 million square kilometers (140 million square miles), or about 71 percent of Earth's total surface area. The major oceans include the Atlantic, Pacific, and Indian oceans. The waters of the marine biomes provide various habitats to more than 250,000 species of organisms. Oceans help regulate climate and weather by contributing to the water cycle and by distributing heat through ocean currents. The oceans also yield an enormous amount of food for humans and are a source for many natural resources. Humans rely on oceans for recreation and for transporting goods throughout the world.

The ocean biome includes several areas or zones. These include estuaries, rocky shores or intertidal zones, the coastal ocean or neritic zone, and the deep ocean zones.

Estuaries

Many major cities have been built on estuaries, including New York City, Tokyo, Buenos Aires, and Rio de Janeiro. Estuaries are coastal bays or inlets where freshwater from a river mixes with salty ocean water. The *salinity* levels of the water also vary throughout the length and depth of the estuaries. Typically, water in the part of the estuary nearest the ocean has a salinity very near that of ocean water. Farther upstream, water is brackish, a mixture of both freshwater and saltwater. The salt content of brackish water is between 1,000 and 4,000 parts per million, but not as salty as ocean water. While the salinity of saltwater is 3.5 percent, (35,000 parts per million), the average salt content in brackish water is generally less than 0.5 percent. Farther up the river, the salinity levels drop.

TABLE 6-2	**Composition of Ocean Water**
Substance	Percent of all Dissolved Material
Chloride	55.0
Sodium	30.6
Sulfate	7.7
Magnesium	3.7
Calcium	1.2
Potassium	1.1
Other	0.7

The salinity of the estuary varies depending on the tides, the amount of freshwater entering from rivers and streams or as rain, and the rate of evaporation. As a result, organisms that inhabit brackish waters are very tolerant of changes in salinity and water depth. Some species avoid periods of high salinity by moving in stream where it is less salty or by burrowing under the bottom sediment.

Several types of rich and fertile wetlands are found in and around estuaries, including salt marshes and mangrove swamps and forests. The characteristics of each wetland are determined by their physical location, the climate of the area in which they are located, and the type of plants that are common to the ecosystem.

SALT MARSHES

Salt marshes are communities whose main producers are salt-tolerant plants that have adapted to saltwater and can survive in fluctuating water levels owing to tide activity. Like all wetlands, salt marshes are transitional areas between land and water. In the freshwater wetlands that occur inland, rainfall and overflow from lakes and rivers are the primary sources of water. Salt marshes, on the other hand, occur along coasts, in bays, lagoons, and other protected coastlines. Salinity and the frequency and extent of flooding determine the types of plants and animals that live in the salt marsh. Salt marshes are places of extreme conditions which change twice daily with the tides.

Instead of trees, the major plants in a salt marsh are grasses, such as the hardy cordgrass. Salt marsh plants possess adaptations that help them survive the very salty conditions of the marsh. Spartina, for instance, is a common marsh grass with stiff, pointed leaves and specialized glands that help it excrete any excess salt absorbed by the roots.

More plants sprout, grow, and die in salt marshes than in almost any other kind of environment. Habitats with high productivity can support many animals because there are lots of plants for them to eat. In a salt marsh, however, cordgrass and other plants are not a direct

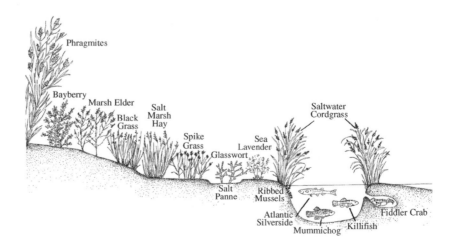

FIGURE 6-7 • **A Salt Marsh Ecosystem**

source of food for the majority of salt marsh animals. Rather, most of the plants are consumed by the process of decay, or decomposition. Decomposition is performed by decomposer organisms, especially microscopic life forms such as bacteria. Scavengers, such as worms, fishes, shrimps, and crabs, feed on the decaying plant material, or detritus, and then excrete the undigested plant remains in feces which can be colonized again by microorganisms. As the microorganisms reduce the detritus into smaller and smaller pieces, it becomes a fertilizer for more growing plants. Daily movement of the tides also enhances the decomposition process. As a result, the entire salt marsh is constantly bathed in a type of fertilizing soup as the nutrients are recirculated.

The rich nutrients provide a habitat for a large diversity of animals and microorganisms. As a rule, the more tropical the climate of the salt marsh, the more diverse its plant and animal life. Among the herbivores that feed on the producers are a variety of zooplankton, mammals, insects, and waterfowl. In addition, many fish live in estuary waters, while amphibians and reptiles make their homes along the coastline. Bottom dwellers include a variety of shellfish, such as oysters, mussels, clams, and barnacles.

Salt marshes perform many functions that are valuable to human beings. They serve as nursery grounds for numerous commercially and recreationally important fish and shellfish species. They act as buffers for the mainland by slowing waves, thereby reducing erosion of the coastline. Furthermore, like all wetlands, they act as filters and help remove sediments and toxins from the water.

Chesapeake Bay

A waterway, located on the mid-Atlantic coast of the United States, Chesapeake Bay contains some of the most important estuaries in North America. This great bay and its watershed encompass more than 165,000 square kilometers (64,000 square miles) along the coasts of New York, New Jersey, Pennsylvania, Delaware, Maryland, Virginia, and the District of Columbia. About 40 rivers, and many times more streams, contribute to Chesapeake Bay, including the Susquehanna, Patuxent, Potomac, Rappahannock, Chester, Choptank, Nanticoke, Wicomico, Pocomoke, York, and James rivers.

The Susquehanna River began to form Chesapeake Bay about 18,000 years ago. As the great continental glaciers over North America began to melt, they caused great rivers of water to flow toward the Atlantic Ocean. The prehistoric Susquehanna, a bigger and more powerful river than its modern descendant, eroded the land as it flowed seaward. As the glaciers melted, the sea level rose, flooding the coast and forming the great bay.

The estuaries along the Chesapeake's shores contain many saltwater and freshwater wetlands of great economic and ecological importance. Significant fish species associated with the Chesapeake Bay include oysters, blue crabs, crayfish, striped bass, speckled trout, shad, and white perch. Annual seafood harvests from Chesapeake Bay may total greater than 45 million kilograms (100 million pounds). These species, and many others, breed or feed in the bay. Some types of development, such as the construction of dams and roads, have hampered the ability of the Chesapeake's freshwater-breeding marine (anadramous) fish, such as shad and herring, to travel upstream to their established spawning grounds far inland.

ENVIRONMENTAL CONCERNS OF SALT MARSHES One of the most important threats to salt marshes, as well as to most other marine ecosystems, is runoff from diverse land locations such as bridges and roads (petroleum products from cars) and from farms and lawns (pesticides and fertilizers). Laws, such as the Clean Water Act (CWA), regulate point-source pollution from an industrial plant, boat, or other single source. Nonpoint-source pollution, such as runoff, is more difficult to monitor and control owing to its multiple sources. Pollution disrupts the food web in the salt marsh by killing off some species and causing others to increase greatly in numbers.

MANGROVES

A mangrove swamp or forest is a biological community that occurs in estuaries along coastlines in tropical and subtropical areas. Worldwide, more than 100 countries have mangrove forests. The most extensive ones are found in Indonesia, Nigeria, Australia, Mexico, Malaysia, and Brazil. The largest mangrove forest on Earth is located in Bangladesh. In the United States, mangrove forests are located at the tip of Florida on the Gulf of Mexico.

Mangrove plants are some of the most remarkable plants in the world. The dominant plant forms of such estuaries are mangrove trees, a type of tropical evergreen. Mangrove trees are adapted to grow in the brackish water conditions of lagoons, swamps, and creeks. They form special stilt-like roots from their branches to breathe in oxygen from the air. This adaptation enables the trees to live in water-logged soil deprived of oxygen. Their roots trap silt and mud, creating a more solid, drier environment over time.

Mangrove swamps and forests provide habitat to a tremendous variety of organisms both above and below the water's surface. Many species of crocodiles, alligators, and small mammals live in mangrove forests. Monkeys, snakes, and a huge variety of birds are found in the canopy—the surface of the ground which is covered by leaves of a tree. The aquatic roots of mangroves support

Manatees, a group of plant-eating marine mammals, live in mangrove swamps where they feed on sea grass. Manatees are on the threatened and endangered list. The three species of manatees are the Amazonian manatee found in the Amazon River, the West African manatee found in the rivers and coastal areas of West Africa, and the West Indian manatee which lives in the Caribbean Sea and along the east coasts of tropical North America (Florida) and South America. Manatees also live in rivers, bays, and estuaries.

FIGURE 6-8 • More than 100 countries have mangrove forests but the most extensive ones are found in Bangladesh, Indonesia, Nigeria, Australia, Mexico, Malaysia, and Brazil.

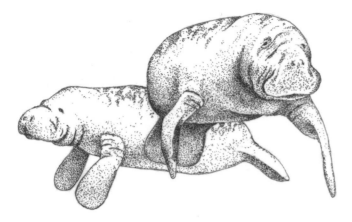

FIGURE 6-9 • Manatees are protected at the federal level by the Marine Mammal Protection Act of 1972 and the Endangered Species Act of 1973.

oysters and barnacles, as well as many fishes, snails, and crustaceans. The mud that collects within mangrove roots also provides a prime habitat for fiddler crabs, various clams, and snails.

ENVIRONMENTAL CONCERNS OF MANGROVES Estimates indicate that more than 50 percent of the world's mangroves have been destroyed. The major causes of the losses can be traced to human activities such as timber harvesting, the conversion of mangroves to aquaculture pools in southeast Asia, and the clearing of the forests for farming activities and housing developments. Mangroves are also subject to excess buildup of sedimentation caused by soil erosion from nearby cleared land and runoff from farms.

Rocky Shores or Intertidal Zones

The rocky shores, also known as the intertidal zones, stretch from the high tide to the low tide marks. Tides are energy forces caused by the periodic rise and fall of ocean waters as well as in some large lakes and inland seas as a result of the gravitational pull of the sun and moon on Earth. Low tide and high tide alternate in a continuous and predictable cycle. Along most coastlines throughout the world, two high tides, or flow tides, and two low tides, or ebb tides, occur each day.

LIFE IN THE INTERTIDAL ZONE

The rising and falling action of the tides creates a zone of plant and animal communities. The constant movement of water transports nutrients into and out of the intertidal zone. However, the pounding of the waves in the intertidal zone also provides a harsh habitat for the organisms that live on the beach, tide pools, and rocky cliffs. Red algae and brown algae are found at the waterline. Microscopic algae grow on the wet sandy beaches and on rock surfaces. During periods of low tides, mollusks and crabs are common residents of tide pools. The upper levels of the rocky cliffs are habitats for periwinkles which can survive for extended periods out of the water. The periwinkles feed on

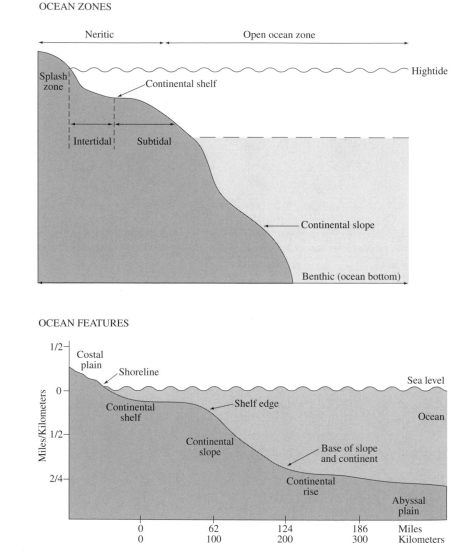

FIGURE 6-10 • Ocean Zones and Features

the algae that grow on the rocky surfaces. Some other living things that live in the intertidal zone include sea urchins, starfish, loons, gulls, and anemones.

Coastal Ocean or the Neritic Zone

The neritic zone comprises the area from a low-tide marking to the outer edge of the continental shelf. The continental shelf is the portion of the ocean floor located between the shoreline and the continental slope. This is an ocean area where the sea floor angles away from the edge of a continent. The continental shelf extends about 160 kilometers (100 miles) offshore and reaches a depth of about 200 meters (600 feet). Here, the shallow, warmer waters and the large amounts of nutrients washed in from the land promote the growth of phytoplankton

Bay of Fundy

The Bay of Fundy is a large intertidal inlet of the Atlantic Ocean located in northeastern Canada, between New Brunswick and Nova Scotia. The funnel-shaped Bay of Fundy is 270 kilometers (168 miles) long and 80 kilometers (50 miles) wide at its mouth. At its head, it forms two basins, Chignecto Basin and Minas Basin. The Bay of Fundy is famous for its tidal bore and for its high tides which reach from 12 to 15 meters (40 to 50 feet), the world's highest tides. It is known as one of the "Marine Wonders of the World."

The Bay of Fundy is a marine ecosystem noted for more than 800 species of animals, including 100 species of fish. The area is critically important as a migratory staging area for millions of birds; it is also a significant summering and wintering area. In any given year, 40 to 70 percent of the world's population of semi-palmated sandpipers stage there. The endangered right whale uses the mouth of the bay as a nursery area for mother-calf pairs and juveniles. In October, the female right whales and their calves swim into the Bay of Fundy where they stay until the end of the month. During this time, the calves learn from their mothers how to swim and dive for food. Sei, fin, minke and humpback whales, and harbor porpoises are also found in the bay from June to October.

The Bay of Fundy is famous for its tidal bore and for high tides that reach 12 to 15 meters. The photo on the left was taken at high tide, and the photo on the right was taken at low tide. (Courtesy of Bay of Fundy.com and Fundy Forum)

and other photosynthetic organisms which trap the energy in sunlight to make food. In turn, this large amount of food attracts a variety of animal species, including many types of fishes and birds, sea otters, dolphins, and porpoises. Other species include sea lions, turtles, penguins, gulls, shellfish, and crustaceans. The continental shelf, the neritic zone, is the most productive area of the ocean for commercial fishing.

Kelp live in the neritic zone in the cold waters along rocky coasts as well as in the intertidal zone. Kelp are the largest, most complex brown algae. Many brown algae have structures called air bladders that help them float near the water's surface, where they are exposed to the sunlight needed for photosynthesis. The body of the seaweed lacks leaves, stems, and roots, but it has structures that anchor the algae to rocks and the ocean bottom. Large kelps may grow to be more than

60 meters (197 feet) in length. Off the California coast, giant kelps form dense underwater forests. These underwater forests provide a rich habitat for a wide variety of marine species.

CORAL REEFS

In the coastal ocean, coral reefs are found worldwide in tropical and subtropical areas that are approximately 25° north and 25° south of the Equator. Coral reefs, composed of calcium carbonate, grow in shallow, tropical water along the coasts of 110 countries. Over one-half of them are located in the Indian Ocean and the Western Pacific area. Another large area of coral reefs is located in the western Atlantic and Caribbean region. The coral growing near the coast are know as fringe reefs; those farther out in the ocean are called barrier reefs. Ring-shaped coral islands that enclose a shallow body of water, or lagoon, are found in the South Pacific, the Indian Ocean, and the Caribbean.

The world's largest coral site is Australia's Great Barrier Reef, which extends more than 1,600 kilometers (1,000 miles) along the northeast coast of Australia. This huge reef is home to different species of coral, birds, and fish. To protect the ecosystem, the Australian government made it a marine reserve in 1975. It encompasses more than 2,500 coral reefs.

In the continental United States, the most extensive living coral are located in the Florida reefs. The largest reefs, off Key Largo, have been designated a marine sanctuary by the National Oceanic and Atmospheric Administration (NOAA). The Key Largo reefs exist at the northernmost fringe of coral development in the Caribbean.

Coral reefs may host more numerous and diverse plant and animal species than any other ecosystem. About 25 percent of all marine species and 20 percent of all marine fish species live in the coral reef

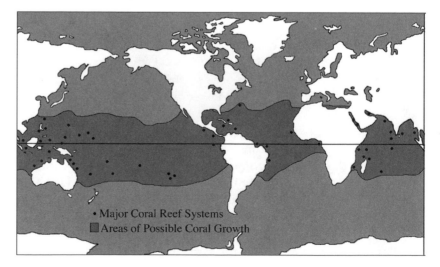

FIGURE 6-11 • The world's largest coral site is Australia's Great Barrier Reef, which extends more than 1,600 kilometers along the northeast coast of Australia.

• Major Coral Reef Systems
■ Areas of Possible Coral Growth

Healthy coral are found in relatively warm and shallow waters in the tropics. (Courtesy of Craig Quirolo)

Coral bleaching is caused when coral lose color and die. Coral bleaching is the result of warming ocean waters thought to be associated with global climate change. (Courtesy of Craig Quirolo)

ecosystem. The reefs are important food sources. Studies indicate that as many as 25 percent of the fish harvested in developing countries are taken from coral reefs. Coral reefs also protect coastlines from ocean storms. The porous structure is ideal for absorbing and dissipating the energy of strong storm waves approaching coastal land areas.

Ocean corals and dinoflagellates such as *zooxanthellae* (golden brown algae) live together in a symbiotic partnership. The algae grow in the cells of the corals. Wastes produced by the corals are used as nutrients by the dinoflagellates. In turn the dinoflagellates provide oxygen and food for the coral through photosynthesis and enhance the rate at which corals build their skeletons. It also provides the coral with its color.

The coral and algae grow upward and over each other during the building process. The coral secrete a limy skeleton that is the basic structure of the reef. New colonies and new coral structures are constantly being built on top of the dead skeletons of older colonies, but the rate of growth is slow. Some corals may grow only about 7.5 centimeters (3 inches) a year. Coral reef building requires warm, well-lit tropical water free from pollution and stirred-up sediments. Extreme water temperatures, dim light, and salinity can cause coral to die.

ENVIRONMENTAL CONCERNS OF CORAL REEFS A study released in 1999, by the World Resources Institute, stated that nearly 60 percent of Earth's living coral reefs are threatened by human activity, including coastal development, overfishing, and inland pollution. Environmentalists are concerned that oil and gas exploration on the continental shelf, along with leaks and oil spills associated with offshore oil drilling, will be hazardous to marine habitats. Mining coral for building materials and natural disasters such as hurricanes can also damage reefs.

Coral bleaching is a phenomenon that causes coral to lose color and die. Some scientists believe that this problem occurs when the dinoflagellates die as a result of abnormally high water temperatures caused by global warming. Highly sensitive corals can live only in water between 18°C and 30°C (64°F and 86°F). Bleaching has occurred when water temperature has increased by 1°C or more. When water temperature increases, the oxygen-deprived coral becomes brittle and stunted. Massive bleaching of coral reefs occurred during 1983, 1987, and 1991 in the Pacific, Caribbean, and Indian oceans. Devastating coral bleaching was recorded in 1998 in regions including Australia, India's Bay of Bengal, the Gulf of Thailand, Florida, and the Seychelles islands off East Africa. However, more conclusive research of warming cycles will need to be done to find out more about coral bleaching.

Open Ocean Zone

The open ocean includes the area from the edge of the continental shelf outward to the deep ocean. It includes the open waters of much of the oceans.

SURFACE ZONE

In the open ocean, sunlight can penetrate only to about 180 meters (590 feet). Life is concentrated close to the surface. Floating at the surface are the producers in the oceanic zone ecosystem, the phytoplankton. The abundant phytoplankton attracts many smaller animals which in turn draw larger predators, such as fishes, dolphins, sharks, and killer whales. There is much less biodiversity in the open ocean than in the neritic zone.

FIGURE 6-12 • The Humpback (top), Right Whale (middle) and Bowhead (bottom) are part of a group of marine mammals of the order *Cetacea* that have flippers and tails with horizontal flukes.

DEEP ZONE OR BENTHIC ZONE

The steep incline at the edge of the continental shelf is the continental slope. The continental slope and the deepest parts of the ocean floor make up the deep, or benthic, zone. Many of the organisms that live in the benthic zone are classified as benthos. The benthic zone is sparsely populated by bottom dwellers, owing to the very cold temperatures and overall lack of producers.

Some of the organisms that live in the deep, dark, benthic zone include sea cucumbers, shrimp, clams, sea stars, lobsters, and sea urchins. These organisms are scavengers and decomposers which rely on the constant downward flow of decaying organic matter from the surface. Because sunlight does not penetrate into the benthic zone, some deep-dwelling species possess special light-generating organs for attracting live food. The angler fish, for example, dangles a luminous lure over its forehead. When small animals bite at the lure, the angler fish swallows them whole. Another species known as the viper fish has luminous structures in its mouth and lives at depths of 600 meters (1,800 feet) or more. The viper fish swims with its mouth wide open so that small prey swim right into the fish's mouth. Some fishes, such as rays, skates, and flounders partially bury themselves on the floor—an adaptation

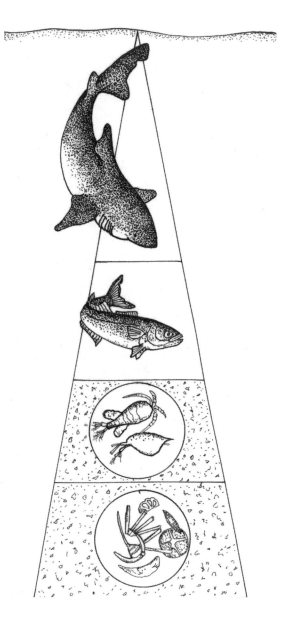

FIGURE 6-13 • A typical ocean surface zone food chain consists of producers and predators.

that provides camouflage as the fish await prey. Other benthic animals, such as sponges, sea anemones, and barnacles, are *sessile* and live attached to surfaces, including those provided by other animals.

Scientists have also discovered rich ecosystems thriving near hydrothermal vents, or places on the ocean floor where hot water and gas flow out of Earth's crust. Growing on rocks in this warm, fertile environment are clumps of bacteria which make energy from the hydrogen sulfide that seeps through the vents. Feeding upon these bacteria are a number of consumers, including giant clams, crabs, and tube worms.

The Viper fish lives at depths of 600 meters and uses its open mouth to catch prey. (Courtesy of Paul H. Yancey, Whitman College, Walla Walla, Washington, USA)

Vocabulary

Benthic zone Bottom of a body of water (freshwater and saltwater) inhabited by decomposers and other bottom dwellers such as clams and insect larvae.

Eutrophication A process by which water is saturated with phosphates and other nutrients which causes an increase in algae growth and destroys other organisms.

Floodplain A broad flat valley through which a river flows.

Littoral A region or zone located between land and a lake or ocean.

Marine Referring to the ocean or sea.

Pelagic zone An area usually in the middle of a body of water where animals live suspended in water.

Salinity The amount of salt dissolved in water.

Sessile Attached directly to a branch or stem.

Zooxanthellae Algae that live in the digestive walls of coral. The zooxanthellae provide coral with oxygen and its color.

Activities for Students

1. Every state in the United States has wetlands. Visit the EPA's Office of Wetlands website and learn about wetlands in your area, how they are at risk, and how you can become involved in conservation efforts.

2. Water biomes have been crucial to the development of human societies. Locate some of the world's major cities on a map. What kind of water features are they near? How did this location help the cities grow into thriving centers of civilization?

3. Visit a pet store and browse through their aquatic animal section. From what water biomes do most animals come? What differences can you spot between freshwater and saltwater species?

Books and Other Reading Materials

Batzer, Darol, ed., et al. *Invertebrates in Freshwater Wetlands of North America: Ecology and Management.* New York: John Wiley & Sons, 1999.

Davidson, Osha Gray. *The Enchanted Braid: Coming to Terms with Nature on the Coral Reef.* New York: John Wiley & Sons, 1998.

Earle, Sylvia A., and Ellen J. Prager. *The Oceans.* New York: McGraw-Hill, 2001.

Kunzig, Robert. *Mapping the Deep: The Extraordinary Story of Ocean Science.* New York: W. W. Norton, 2000.

Longhurst, Alan R. *Ecological Geography of the Sea.* San Diego, Calif.: Academic Press, 1998.

Niering, William A. *Wetlands.* Audubon Society Nature Guides. New York: Knopf, 1985.

Sale, Peter F., ed. *The Ecology of Fishes on Coral Reefs.* San Diego, Calif.: Academic Press, 1994.

Websites

American Oceans Campaign, http://www.americanoceans.org

American Rivers, http://www.amrivers.org

Chesapeake Bay and Marshes, http://www.fwdj.com/chesape.html

Enviromental Literacy Council, http://www.enviroliteracy.org

Environmental Protection Agency, Office of Wetlands, Oceans and Watersheds, National Wetlands, http://www.epa.gov/owow

Institute of Cetacean Research (ICR), whales, http://www.whalesci.org

National Alliance of River, Sound, and Bay Keepers, http://www.keeper.org

National American Association for Environmental Education, http://naaee.org/npeee/npeee.html, for activities, see the Environmental Education Collection, a review of resources for educators, vols. 1, 2, 3

National Estuary Program, http://earth1.epa.gove/nep/

National Marine Mammal Laboratory, http://nmml01.afsc.noaa.gov

National Oceanic and Atmospheric Administration, Coastal Zone Management Program, http://www.nos.noaa.gov/ocrm/czm

National Wetlands Research Center, http://www.nwrc.usgs.gov/educ_out.html

Ocean Planet, http://seawifs.gsfc.nasa

Ramsar Convention on Wetlands (International), http://www2.iucn.org/themes/ramsar/

Reef Relief, http://www.reefrelief.org

Save the Manatee Club, http://www.seaworld.org/manatee/sciclassman.html

USGS Coastal and Marine Geology, salt marshes, http://marine.usgs.gov/

CHAPTER 7

Changes in the Ecosystem

Mount Saint Helens, an active volcano, is located in the Cascade Mountains of Washington State. On May 18, 1980, a major eruption occurred on the slopes, and much of the northern side of the volcano was blown away. Thousands of trees in the forest were knocked down and flattened by *landslides*. When it was over, more than 50 people had lost their lives. About 400 square kilometers (150 square miles) of forest was covered with volcanic rock, mud, and ashes. On that day, the entire ecosystem on Mount Saint Helens was destroyed.

Within a short time after the destruction, however, animal and plant life began to be reestablished on the slopes. Today, the forest is slowly coming back. The forest is regenerating. Shrubs and small trees are now growing on the slopes, and even some animals have returned.

The destruction of the Mount Saint Helens ecosystem illustrates that all ecosystems are open to the forces and events of change. Some of the events that can cause changes in an ecosystem include volcanoes, earthquakes, fires, floods, droughts, climate changes, and human activity. Evolution and extinction also effect changes in ecosystems.

IMPACT OF NATURAL DISASTERS ON ECOSYSTEMS

Natural disasters are geological or weather-related events, such as forest fires, floods, and earthquakes, which kill and injure people and other organisms, destroy personal property, damage the environment, and

FIGURE 7-1 • Mount Saint Helens is located in the Cascade Mountains of Washington State. The summit height is 2550 meters. Before the volcano erupted, it was 3200 meters high.

disrupt a country's economy. Natural disasters can occur at any time on any continent.

Floods and Forest Fires

Floods, the most common type of natural disaster, usually occur when heavy rains cause rivers and streams or lakes to overflow. Floods can cause extensive destruction to people, homes, automobiles, buildings, and other structures by submerging them under water. In addition, floods may disrupt ecosystems by destroying the habitats of organisms or changing the conditions of the habitat.

Forest fires most often occur after periods of drought when environmental conditions are very dry. Fires often devastate ecosystems by burning the trees and other vegetation that provide habitat or food for animals and other organisms. Such fires also release pollutants such as carbon dioxide, soot, and ash into the atmosphere. While the damaging effects of forest fires are clear, there are also some ecological benefits of fire. For example, the cones of the Jack pine, a common conifer in North America, release seeds only after they have been exposed to the heat from a forest fire. Fires also help clear dead wood, brush, and other accumulated debris that might otherwise lead to more fires.

Earthquakes and Volcanoes

Earthquakes occur when the tectonic plates that make up Earth's crust move and scrape against each other. The movements can occur along a *fault*. However, the movements of the crustal plates are themselves very slight. However, they are significant enough to send vibrations up through the crust, causing the surface to shake, often violently. Earthquakes are not particularly damaging to ecosystems, but they can be

Although forest fires can be devastating to the environment, some fires can also serve ecosystems by recycling nutrients, regulating plant succession and wildlife habitat, reducing biomass, enriching soils, and controlling insect and disease organism populations. (Courtesy of National Interagency Fire Center, Bureau of Land Management)

devastating to people and their possessions. Strong earthquakes can collapse buildings, roads, bridges, and other structures.

Refer to Chapter 1 for a description of tectonic plates.

Each year between 25,000 and 50,000 earthquakes are reported. Most of them are minor and cause no damage; however, some are powerful and damaging earthquakes. In 1999 an earthquake in Turkey claimed more than 16,000 lives, and almost 38,000 others were reported injured. Thousands were homeless. The earthquake was one of the most powerful recorded in the twentieth century. There were as many as 250 aftershocks. In 1995 a powerful earthquake rocked Osaka, Kyoto, and Kobe, in Japan. More than 5,000 people were killed, 26,000 were injured, and the property damage was about 100 billion dollars.

A volcano is an opening in the Earth's surface through which molten rock, called *lava*, and gases are released into the environment. Today, there are about 600 active volcanoes around the world. Volcanic eruptions are impressive, yet destructive, events. People are often not greatly affected by volcanic eruptions because cities and towns tend not to be built too near active volcanoes. Nevertheless, damage to the environment within the vicinity of a volcano can be quite extensive.

Much of the activity of earthquakes and volcanoes is located around the edge or rim of the Pacific Ocean—an area extending off the eastern section of Asia and along the western edge of North and South America.

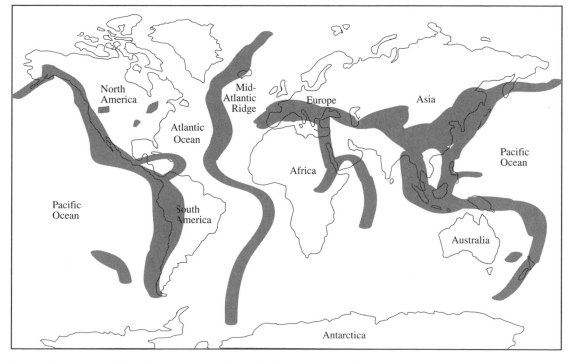

FIGURE 7-2 • Much of the activity of earthquakes and volcanoes is located around the edge or rim of the Pacific Ocean—an area extending off the eastern section of Asia and along the western edge of North and South America. An earthquake in the Iranian city of Bam accounted for 41,000 deaths in 2003.

The zone is known as the "Ring of Fire" and the Circum–Pacific belt. This is a system of subduction zones on the edge of the Pacific plate. About 90 percent of all the Earth's earthquakes occur in this zone. Another major zone of active areas of earthquakes extends from the Mediterranean region, eastward through Turkey, Iran, and northern India. Earthquakes can occur in other areas of the world as well.

Scientists use the Richter scale to measure the intensity of seismic events. The scale, developed by Charles Richter in the 1940s, measures an earthquake's intensity or magnitude on a scale of 1 to 9. A strong earthquake, such as the one recorded in Chile in 1960, measured 8.3 on the Richter scale. The Richter scale measures the intensity of seismic events without respect to damage.

The method used by seismologists to assess the intensity of damage of an earthquake is known as the modified Mercalli scale. The scale uses Roman numerals I to XII to designate the degree of damage: I indicates an earthquake that is not felt except under unusual conditions, whereas XII indicates total destruction.

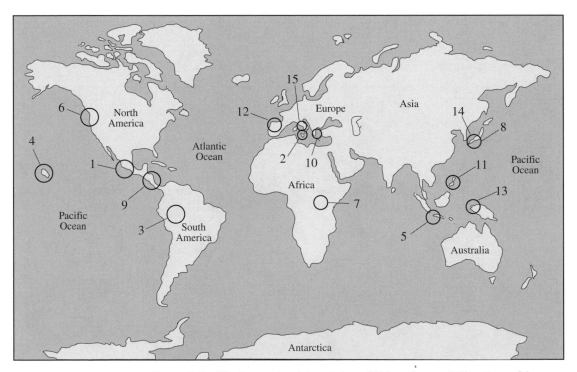

FIGURE 7-3 • The International Association of Volcanology and Chemistry of the Earth's Interior, in cooperation with the United Nations' International Decade for Natural Disaster Reduction, are studying these major and potential dangerous volcanoes. All of them are near heavy populated areas. *Source:* United Nations' International Decade for Natural Disaster Reduction. (Key to map: 1. Colima, Mexico; 2. Etna, Sicily; 3. Galera, Colombia; 4. Mauna Loa, Hawaii; 5. Merapi, Indonesia; 6. Mount Rainier, Washington; 7. Nyiragongo, Zaire; 8. Sakura-Jima, Japan; 9. Santa María, Guatemala; 10. Santoríni, Greece; 11. Taal, Philippines; 12. Teide, Spain; 13. Ulawun, New Guinea; 14. Unzen, Japan; 15. Vesuvius, Italy)

TABLE 7-1	Earthquake Magnitudes	
Year	**Location**	**Richter Number**
1811–1812	New Madrid earthquake, Missouri (a series of quakes from December 1811 to March 1812)	8.0–8.3
1899	Yakutat Bay earthquake, Alaska	8.3–8.6 (estimate)
1906	San Francisco earthquake, California	7.7–8.25 (estimate)
1964	Alaska earthquake, Alaska	8.5
1976	Tangshan earthquake, China	8.2
1985	Mexico City earthquake, Mexico	8.1
1990	Northwest Iran earthquake, Iran	7.7
1994	Bolivian earthquake, South America	8.3 (largest deep quake on record)
1995	Kobe earthquake, Japan	6.8

Hurricanes, Typhoons, and Monsoons

Coastal hazards include hurricanes and monsoons. A hurricane is a violent tropical cyclone that occurs in the western Atlantic Ocean and Caribbean regions. A cyclone is an area of low pressure around which the air turns in the same direction as Earth.

HURRICANES

Hurricanes have sustained wind speeds of at least 150 kilometers (90 miles) per hour. The high winds associated with hurricanes cause huge waves that batter coastal areas, leading to flash floods, beach erosion, and often significant damage to marinas and shoreline homes. When hurricanes move inland, their powerful winds and torrential rains can destroy homes, buildings, and power lines and cause extensive damage to crops and natural resources. Drowning is the major cause of death resulting from hurricanes. Some of the most disastrous hurricanes in the 1990s have included the following:

- In 1992 Hurricane Andrew swept along a path through the northwestern Bahamas, the southern Florida peninsula, and south-central Louisiana. Damage was estimated to be near 25 billion dollars, making Andrew the most expensive natural disaster in the history of the United States as of that date.

- In 1994 Hurricane Gordon followed an unusual, erratic path from the western Caribbean Sea and islands and then to Florida and the southwestern Atlantic Ocean. Its torrential rains caused much loss of life in Haiti and extensive agricultural damage in southern Florida.

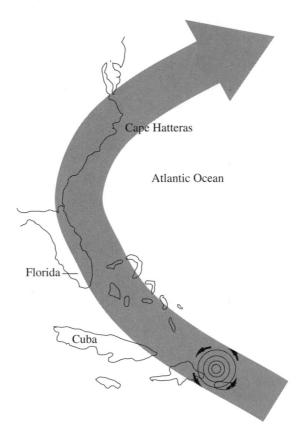

FIGURE 7-4 • The Path of a Typical Hurricane
Hurricanes form in all of Earth's tropical oceans except in the cool waters of the South Pacific and South Atlantic Oceans.

- In 1998 Hurricane Mitch brought massive devastation and loss of life across Central America. Officials estimate that Mitch left more than 10,000 people dead across Central America. An additional 13,000 people were missing and presumed dead, and a staggering 2.8 million people were left homeless. Hardest hit were Honduras and Nicaragua.

TYPHOONS

A typhoon is a severe tropical cyclone that occurs in the western Pacific Ocean region in areas between 5° and 30° latitude. As severe storms, typhoons can cause great damage to physical structures and property, be devastating to organisms, and bring about both short-term and long-term changes to the environment.

MONSOONS

A monsoon is a major wind system that changes its direction during a seasonal change. During one season, the winds blow from sea to land. During another season, there is a complete reversal of wind direction, and the winds blow from land to sea. The monsoon is caused by differences in annual temperature and air pressure between land and sea.

El Niño/La Niña

A phenomenon called El Niño (pronounced el NEEN-yo) is a major climatic event which occurs periodically in the eastern Pacific Ocean and produces remarkable weather changes over large portions of Earth. The National Oceanic and Atmospheric Agency describes El Niño as a disruption of the ocean-atmosphere system in the tropical Pacific with important consequences for weather around the globe.

The El Niño climate cycle begins in an area of warm water in the eastern Pacific Ocean where high atmospheric pressure causes the prevailing trade winds to blow to the west. These west-bound trade winds result in a higher sea level in Indonesia. Rainfall is found in the rising air over the warm water in the western Pacific with little or no rainfall in the eastern Pacific. Due to the cooler temperatures in the eastern Pacific, there is an upwelling of cold water from the bottom of the ocean which is rich with nutrients supporting many diverse marine ecosystems and fisheries. However, an occasional slackening of the trade winds will cause the usual westward movement of water to cease and cause, instead, an eastward flow of water to occur beneath the surface of the ocean. This eastward flow depresses the thermocline, or the layer of cold water. As a result, the eastern Pacific ocean temperatures rise to greater temperatures than usual at the surface. Rainfall follows the eastward movement of the warm surface waters causing flooding conditions in Peru. Due to the lack of rain in the west, dry conditions occur in Indonesia.

La Niña ("the little girl," or sometimes known as "El Viejo," or "the old man"), the counterpart of El Niño, occurs at the opposite point in the climatic cycle. The El Niño–La Niña cycle consists of a continuously changing balance between two extremes of climatic conditions: as El Niño fades, La Niña emerges.

Weather impacts attributable to El Niño during 1998 included flood damage in Southern California, a mild hurricane season in the Caribbean Sea, severe drought in Australia, and forest fires in Indonesia. During the winter, snow fell in the southwestern United States and in Mexico in much greater quantities than usual. The west coast of South America suffered from heavy rains and snow melt, with floods in Peru and mudslides in Ecuador.

The 1998 El Niño episode caused the deaths of thousands of southern sea lions and South American fur seals along the coasts of Chile, Peru, Ecuador, and the Galapagos. These marine mammals died from starvation. Important food species such as anchovies and sardines became scarce owing to the migration of the fish away from the warmer-than-usual coastal waters. The sea lions and seals were forced to travel much farther from their home territories in search of food and a large proportion of infant animals died.

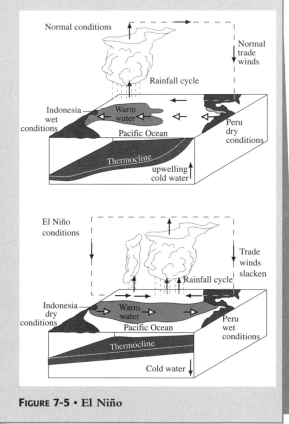

FIGURE 7-5 · El Niño

Monsoons appear along the Gulf Coast of the United States and in central Europe, but the best developed ones occur in South Africa, northern Australia, southeast Asia, and, particularly, India.

In India, for example, there are two distinct monsoon seasons. In the summer, the monsoons move over the Indian Ocean and blow southwest over the land. The summer monsoons bring heavy winds and significant precipitation in the form of rain. This is an important growing season. In the winter, monsoons blow across the land in a northeast direction out to the Indian Ocean. Winter monsoons, by contrast, are cold and dry with very little precipitation. The dry, harsh winds can cause droughts and damage or destroy crops. The summer monsoons can also pose problems. Any disruption of the summer monsoon can be a disaster to heavily populated towns and cities in the monsoon belt. Too much rain can cause massive flooding conditions, particularly in low-lying coastal settlements, and too little rain can impact farmers who depend on the rain to grow crops during the growing season.

When natural disasters occur, such as those mentioned, scientists try to learn from them so they can identify the factors that cause them. Instruments, such as weather satellites, computers, and ground sensors are used to gather data about conditions in the atmosphere and on Earth. Modern technologies such as these enable scientists to predict with more accuracy when natural disasters may strike.

ECOLOGICAL SUCCESSION

After an ecosystem has been destroyed, it can be replaced through a process known as ecological succession, which is the natural process through which one kind of community succeeds another. Neglecting a vegetable garden, for example, will invite weeds and other plants to grow in place of the vegetables over a period of time. Constant garden care is therefore needed to keep out undesirable plants.

Succession occurs partly in response to changes in the abiotic, or physical, characteristics of an ecosystem and partly as a result of changes in the types of organisms that make up the producer level of that ecosystem. There are two types of ecological succession: primary succession and secondary succession.

Primary Succession

Primary succession occurs when a variety of successive communities colonize an area that was not previously inhabited by organisms. The first stage in primary succession is the development of soil. Primary succession occurs in areas that lack the soil needed for plants (the main producers of terrestrial ecosystems) to colonize the area. Areas in

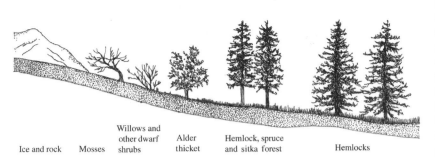

Ice and rock Mosses Willows and other dwarf shrubs Alder thicket Hemlock, spruce and sitka forest Hemlocks

FIGURE 7-6 • Primary succession occurs when a variety of successive communities colonize an area that was not previously inhabited by organisms. This diagram illustrates the primary succession stages that may occur after a retreating glacier.

which primary succession is likely to occur include bare rocks recently exposed by a retreating glacier. A lava flow from volcanic action also starts primary succession once the lava field cools. Other examples would be newly created volcanic islands.

An example of primary succession occurred on the island of Rakata, once known as Krakatoa. This Pacific island is located off southwestern Indonesia between Sumatra and Java. In 1883 repeated volcanic eruptions destroyed almost half of the small island. The island was approximately 47 square kilometers (18 square miles) in size prior to the eruption of 1883; however, the impact of the explosion was so severe, the island was reduced to its current size of only 16 square kilometers (6 square miles). What was left was covered with volcanic debris—ash and *pumice*. All of the plants and animals on the island were destroyed. In addition, thousands of people living along the coasts of Sumatra and Java were killed by the waves resulting from the *tsunami* caused by the eruption and an accompanying earthquake.

Krakatoa demonstrates how devastating a volcanic eruption can be to the surrounding environment. However, primary succession took over on the krakatoa, and within 20 years more than 100 species had established themselves on the land. In time, some of these species disappeared and were replaced by other, more dominant species.

Secondary Succession

Secondary succession occurs in areas in which other biological communities previously lived but were displaced. This situation could result from a natural disaster or through a major change in the area such as the clearing away of its vegetation.

Mount Saint Helens is a good example of secondary succession. Secondary succession occurs when an established ecological community, such as the forest on Mount Saint Helens, is disturbed. In secondary succession, the presence of soil permits an area to support the growth of small plants such as mosses and grasses. If the seeds of such plants are carried into the area by wind, running water, or passing animals, and conditions for growth are favorable, new plants may begin to grow. Without competition from other organisms, these plants can thrive, allowing a large plant population to develop quickly. If a lichen population were present (as occurs during primary succession), grasses and mosses will replace it.

The mosses and grasses that identify an area at the beginning of secondary succession may thrive for many generations. Several generations of these small plants continue to go through their life cycles. They help build new soil as the plants that die are broken down through physical and chemical processes as well as through the activities of bacteria of decay and other decomposers. In time, the soil becomes deep enough and rich enough to support the growth of larger plants such as shrubs. At the same time, animals may begin to colonize the area as plant sources of food become more available. The shrub and animal populations may thrive for several generations, with the deaths of individuals again helping to build and enrich the soil. In time, this soil too will become able to support the growth of larger plants, allowing small trees to begin growing in the area.

The communities at each level of succession are not very stable. They are easily displaced by newer populations that outcompete the previous populations for resources or by natural disturbances such as fire or flood. In some climates, if succession continues to advance, the shrub and small tree population may in time be replaced by larger trees, whose thick *foliage* prevents sunlight from reaching the ground. Without sunlight, many of the ground-level plants will die off and be replaced by larger trees. In turn, the animal populations of the ecosystem also change as new and diverse habitats become available and the major plant food source in the area changes.

The first trees in a community to colonize an area often are coniferous trees, whose roots are shallow and spread out laterally near the surface of the soil. In time, as the soil becomes deeper, these trees may be replaced by a community comprising broadleaf deciduous trees such as birch, aspen, and maples mixed with larger conifers. As a result, a dense forest ecosystem will develop. The forest ecosystem is generally more stable than the earlier communities that inhabited the area. If left undisturbed, this community will thrive in the area and not undergo further succession. This final, stable community is referred to as a climax community. A climax community exists when no other types of species are able to compete successfully with that community.

Pioneer Species

Pioneer species are the first group of organisms to colonize a barren, lifeless area, thus beginning the process of ecological succession. During primary succession, pioneer species may colonize areas composed largely of bare rock, such as those left exposed by a retreating glacier. However, in secondary succession, pioneer species are the first to colonize areas left barren and lifeless as a result of natural disasters such as

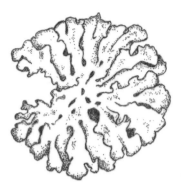

Figure 7-7 • Pioneer species vary according to the type of succession an area is undergoing. In primary succession, such species generally include lichens (left), microorganisms, and a few small, hardy plant species.

floods and earthquakes, or through the activities of humans such as agricultural abandonment or deforestation.

Pioneer species vary according to the type of succession an area is undergoing. In primary succession, such species generally include microorganisms; a few small, hardy plant species; and lichens.

Lichens are often the first organisms to colonize an area that will undergo primary succession. These organisms are a symbiotic association of an alga or cyanobacterium and a fungus. As the organisms making up a lichen carry out their life processes, acids and other substances given off by the organisms tend to break down, through decomposition and weathering, the surfaces of the rocks upon which these organisms live. This breaking down of rock results in the formation of new soil. Lichens are known as a pioneer species because they are the first organisms to colonize an area during primary succession. However, in secondary succession, pioneer species often include plants such as mosses and grasses, particularly those hardy grasses that are generally described as weeds.

Succession in Aquatic Ecosystems

Succession does not occur only in terrestrial environments. For example, lakes and ponds also undergo succession as they slowly fill with sediments that allow plants to begin growing. Soil generally begins to fill in a lake or pond at its edges. In addition, aquatic plants may begin to take root and thrive in the water of the lake or pond. Over many years, the growth of plants and the development of the new soil that results from their activities inches nearer and nearer to the center of the lake or pond, causing the water to become more shallow. In time, the lake or pond community may be replaced by a marsh or swamp.

As the aquatic plants making up a swamp or marsh continue to go through their life cycles, the breakdown of the remains of these plants results in the formation of more soil. Over many years, enough soil may collect in the wetland to allow the aquatic plants to be replaced by grasses and mosses; a terrestrial ecosystem replaces the aquatic ecosystem. As times goes on, the grasses may form a *meadow* that

DID YOU KNOW?

In 1988, a large forest fire burned thousands of acres of trees and bushes in Yellowstone National Park. After the fire, the pioneer species were wildflowers. In the early 1990s, grasses, ferns, and other plants replaced the wildflowers.

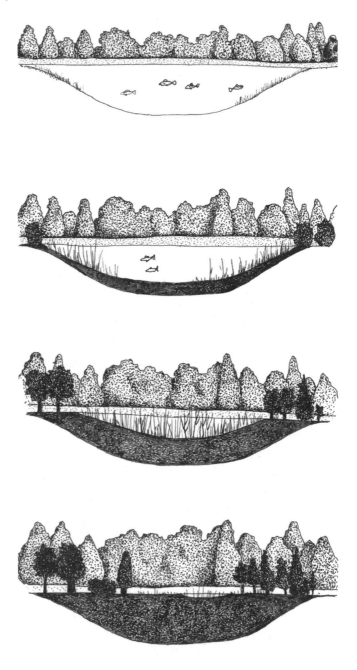

FIGURE 7-8 • **Succession in an Aquatic Ecosystem**

may continue to undergo succession until it is replaced by a more stable forest community.

IMPACT OF EVOLUTION ON ECOSYSTEMS

The process by which the *traits* of populations of organisms change over time is known as evolution. The theory of evolution enables scientists to explain the diversity that exists among Earth's organisms by providing the means through which new species arise from existing species. The theory of evolution accepted by most scientists was first proposed by British naturalist Charles Darwin in 1859. This theory provided not only explained how new species arose but also why organisms are adapted to life in particular environments.

Darwin's theory of evolution contains four main observations. First, all organisms are capable of producing more offspring than can survive. As an example, a single female frog may lay thousands of eggs at one time; however, only a small number of the eggs will actually hatch and survive to develop into adult frogs. Many eggs may not be fertilized or may be eaten by other organisms before they hatch. Similarly, some organisms that do hatch may be eaten by other organisms before they develop into adults.

Second, individual organisms of the same species have differences, or variations, in their traits. Differences in traits may include such things as coloration, running speed, size, and strength, as well as behavioral differences and resistance to disease. Although Darwin did not know it at the time, these differences in traits result from the *genes* inherited by an organism from its parents.

Third, some traits provide the organism with a survival advantage in its environment. Known as natural selection, this hypothesis suggests that organisms having advantageous traits are more likely to survive and reproduce in their environments than are organisms lacking these traits. This process occurs over long periods of many generations.

Fourth, traits providing a species with advantages for survival in its environment will be passed on to the offspring of these organisms. Thus, the offspring will have the same advantages for survival as did their parents. Through the process of evolution, these advantageous traits will become more common in successive generations, resulting in more and more individuals with the advantageous traits. At the same time, traits not selected in earlier generations will become less common in successive generations.

Scientists use Darwin's theory of evolution by natural selection to explain why organisms are so well adapted to their environments and why so many different species exist on Earth. For example, when organisms are not well adapted to their environments, they must move to a different environment to which they are adapted, or they will die. Upon doing so, they will undergo the same process of evolution by natural selection as does the population from which they originated. Over time, the traits of this population may differ enough from those of their ancestral population that a new species emerges.

Adaptive Radiation

Adaptive radiation is an evolutionary process that results in the emergence of new forms of organisms at the species, genus, or family level from a common ancestral population. The process of adaptive radiation is evident in several distinct finch species observed among the Galapagos Islands by Charles Darwin. Each species had differences in its beak and foot structure that enabled it to eat foods and live in habitats distinct from those of other finch species. Despite these differences, Darwin hypothesized that all the finches shared a common ancestor. These observations helped Darwin develop his theory of evolution based on natural selection.

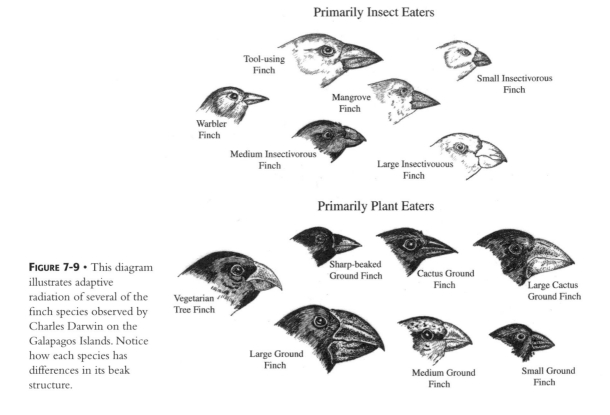

FIGURE 7-9 • This diagram illustrates adaptive radiation of several of the finch species observed by Charles Darwin on the Galapagos Islands. Notice how each species has differences in its beak structure.

IMPACT OF EXTINCTION ON ECOSYSTEMS

DID YOU KNOW?

Scientists use carbon-14 to date fossils less than 50,000 years. For fossils and rocks that are millions and millions of years old, scientists use potassium-40.

Extinction is the disappearance of a species as a result of a natural disaster, climate change, overspecialization, unsuccessful competition for resources, or human-caused stresses such as habitat destruction, hunting, and introduction of exotic species. Extinction has occurred since the beginning of life on Earth about 4.6 billion years ago; in fact, most of the organisms that have ever existed are now extinct. Today's increased concern about extinction is not that it occurs, but that human activities have greatly accelerated the species extinction rate far beyond the natural background rate. Conservative estimates put the current rate of extinction at one species per day. At a minimum, 100,000 species have died out within the past 100 years. Experts predict that if present trends continue, we are likely to lose half of all living species within the next century.

Of all the species that have lived on Earth since life first appeared, only about one in a thousand is still living today. All others have become extinct, typically within about 10 million years of their first appearance. This high extinction rate has had an important influence on the evolution of life on Earth, and on the world's great biodiversity.

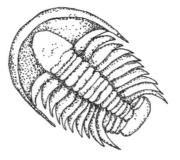

The population and repopulation of an ecological niche by species after species allow for the testing of a wider range of survival strategies than the slower process whereby a species gradually adapts itself to its environment.

Several mass extinctions—sudden disappearances of a large number and variety of species within short periods of time—are recorded in the fossil record. Such large-scale extinctions may have been caused by catastrophic agents, such as meteorite impacts, or terrestrial agents, massive volcanic eruptions, ice ages, sea level rises, global climate changes, and changes in ocean oxygen or salinity levels.

Paleontologists have been able to recognize patterns within and between such extinction events. Mass extinction has struck both marine and terrestrial species. On land, animal species have tended to be the hardest hit, while plant species have been more resistant. There have been frequent disappearance of tropical forms of life during mass extinctions.

Mass extinctions have occurred about every 26 million years; scientists recognize between 5 and 10 major episodes in the geologic record. Dinosaurs were the well-known victims of a major mass extinction that occurred 65 million years ago, at the boundary between the Cretaceous and Tertiary periods. The world's largest mass extinction, however, took place about 240 million years ago, at the end of the Paleozoic era. Scientists estimate that, during that event, between 80 and 96 percent of all species disappeared from Earth.

The overall pattern of extinction within human time, about the past 2 million years, generally follows the movement of humans from Africa and Asia to Europe, and their subsequent migration to the Americas and the world's islands. Prehistoric (circa 11,000 years before the present) North American species extinctions caused, directly or indirectly, by human overhunting include the mammoth, mastodon, woolly rhinoceros, dire wolf, and sabre-tooth cat. Well-known species that have become extinct in historical time, also because of human

Dodo Bird

The dodo bird was a native bird of the island of Mauritius in the Indian Ocean. In the 1500s, trading ships landed on the island and sailors hunted the dodos for food. Later nonnative animals such as rats, pigs, and monkeys were introduced to the island. Many of these animals preyed on the dodos and ate their eggs. The combination of overhunting, destroyed habitats, and the introduction of nonnative animals took its toll on the dodo population. The dodo bird became extinct around the 1680s.

FIGURE 7-11 • The dodo bird was a flightless bird that became extinct. It lived on many islands in the Indian Ocean.

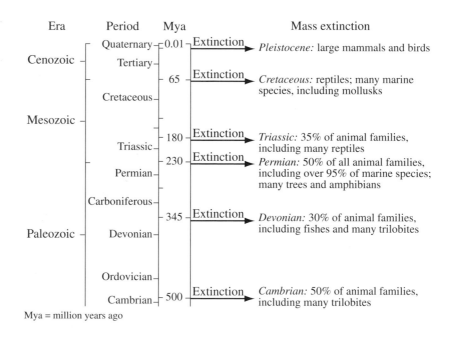

FIGURE 7-12 • Mass Extinction

activities, include Steller's sea cow, the great auk, the Carolina parakeet, and the passenger pigeon.

The many extinctions occurring in the world today involve complex issues that are in many cases relatively new to science and have global implications. A complete inventory of species threatened with extinction would include every type of organism: microorganisms, fungi, plants, and animals. Because we know so little about the variety of life on Earth, it is likely that many species have already died out without our ever knowing of their existence.

One method scientists use to understand and explain extinction is to focus research on an isolated region, such as an island, or on a particular type of organism. Using information gathered from such specific studies, scientists hope to assemble a clearer picture of worldwide extinction patterns.

IMPACT OF HUMANS ON ECOSYSTEMS

The increase in the population of human beings and their activities such as farming, mining, manufacturing, fishing, and hunting has had an impact on regional ecosystems and global biomes. Human activities have resulted in air, land, and water pollution that has damaged, destroyed, and changed ecosystems causing various species to become extinct or threatened with extinction. Pollution has also resulted in major health hazards and health care costs.

Air Pollution

Air pollution is the result of contaminants in the air, from mobile or stationary sources, which degrade natural air quality and often present a health hazard. Outdoor air quality is affected by many human and natural activities. Manufacturing companies, power plants, small businesses, automobiles, forest fires, and volcanoes all are potential sources of air pollution, as is any activity that releases materials into the air. Pollutants in the air can create smog and acid rain, cause respiratory disease and other serious illnesses, damage the protective ozone layer in the upper atmosphere, and contribute to climate change and global warming.

Water Pollution

Water pollution refers to the process and the result of adding pollutants to naturally occurring water, including groundwater and surface water, such as rivers, lakes, and oceans. Water pollution can include chemical pollution, biological pollution with microorganisms, or thermal (heat) pollution, which is the changing of the water temperature. Other sources of water pollution include stormwater runoff, acid mine drainage, acid rain, and leaking underground storage tanks. Thermal pollution can result from discharge of runoff from hot pavement or from industrial cooling water. High concentrations of pesticides, nitrogen compounds, and phosphorus materials, as well as soil erosion, can enter waterways covering streambeds, decreasing the oxygen content of the water, and reducing the ability of plants and animals to support themselves. The pollution impacts water quality and can threaten drinking water supplies. Air pollution and water pollution cost citizens billions of dollars every year in health care costs and lost productivity.

Vocabulary

Fault Line of a crack in Earth's crust where the shifting of crustal plates takes place.

Foliage Leaves of a plant.

Gene Specific part of DNA (deoxyribonucleic acid) located on a chromosome.

Landslide Sudden fall of large amounts of soil and rock down the side of a hill or mountain.

Lava Molten material which flows from an erupting volcano.

Meadow Field of grass and wild plants.

Pumice Light glass-like substance formed at the edge of a lava flow.

Trait Characteristic particular to an organism.

Tsunami Tidal wave caused by an earthquake under the ocean which moves rapidly across the ocean's surface.

Activities for Students

1. Learn about the eruption in AD 79 of Mount Vesuvius in Pompeii, Italy. How much destruction did it cause? The area around this active volcano has been repopulated. What advice might you give to the people who live near its base?

2. What animals are currently in danger of becoming extinct? Go to your local zoo and visit the animals that are at risk. What regions of the world do they come from, and what is happening in those places that creates a threat to these creatures?

3. Air pollution is a hazard faced by all people who live in or near urban areas. Start an education campaign in your community on the problems associated with pollution and brainstorm ways in which people can help reduce the toxins in the air.

Books and Other Reading Materials

Bernstein, Leonard, Alan Winkler, and Linda Zierdt-Warshaw. *Environmental Science: Ecology and Human Impact.* Menlo Park, Calif.: Addison-Wesley, 1995.

Bolt, Bruce A. *Earthquakes.* New York: W. H. Freeman, 1993.

Elsneer, J. B. *Hurricanes of the North Atlantic: Climate and Society.* Oxford, England: Oxford University Press, 1999.

Facklam, Margery, and Pamela Johnson. *And Then There Was One: The Mysteries of Extinction.* Boston: Little, Brown, 1993.

Glantz, Michael H. *Currents of Change: El Niño's Impact on Climate and Society.* Cambridge, England:Cambridge University Press, 1996.

Lessem, Don. *Dinosaurs to Dodos: An Encyclopedia of Extinct Animals.* New York: Scholastic Trade, 1999.

Rees, Robin, ed. *The Way Nature Works.* New York: MacMillan, 1992.

Sieh, Kerry E., and Simon LeVay. *The Earth in Turmoil: Earthquakes, Volcanoes, and Their Impact on Humankind.* New York: W. H. Freeman, 1998.

Zimmer, Carl, and Stephen Jay Gould. *Evolution: The Triumph of an Idea.* New York: HarperCollins, 2001.

Websites

Center for Integration of Natural Disaster Information, http://cindi.usgs.gov/events/

Endangered Species Recovery Council, http://www.esrc.org

Exploratorium website, http://www.exploratorium. edu/faultline

International Union for the Conservation of Nature/World Conservation Union, http://www.iucn.org

National Hurricane Center, http://www.nhc.noaa.gov

U.S. Fish and Wildlife Service, Division of Endangered Species, http://www.fws.gov/r9endspp/endspp.html

U.S. Geological Survey Website, http://quake.wr.usgs.gov/; http://geology.usgs.gov/quake.html; http://geohazards.cr.usgs.gov/earthquake.html; http://geohazards.cr.usgs.gov/eq/

World Wildlife Fund, http://www.panda.org

Appendix A: Environmental Timeline, 1620–2004

Environmentalists and activists appear in **boldface**.

1620 to 1860 Erosion becomes a major problem on many American farms. Fields are abandoned. Rivers and streams are filled with silt and mud. The publication of farm journals is initiated by early soil conservationists to improve farming methods.

1748 Jared Eliot, a minister and doctor of Killingsworth, Connecticut, writes the first American book on agriculture to improve crops and to conserve soil.

1824 Solomon and William Drown of Providence, Rhode Island, publish *Farmer's Guide* which discusses erosion and its causes and remedies. A year later, John Lorain, of the Philadelphia Agricultural Society, publishes a book devoted to the prevention of soil erosion in which he discusses methods such as using grass as an erosion-control crop.

1827 John James Audubon begins publication of *Birds of America*.

1830 George Catlin launches his great western painting crusade to document Native American peoples.

<1845 Henry David Thoreau moves to Walden Pond to observe the fauna and flora of Concord, Massachusetts.

1847 U.S. Congressman **George Perkins Marsh** of Vermont delivers a speech calling attention to the destructive impact of human activity on the land.

1849 The U.S. Department of the Interior (DOI) is established.

1857 Frederick Law Olmsted develops the first city park: New York City's Central Park.

1859 British naturalist Charles Darwin publishes *The Origin of the Species by Means of Natural Selection*. In time the theory of evolution presented in the book becomes the most widely accepted theory of evolution.

1866 German biologist Ernst Haeckel introduces the term *ecology*.

1869 John Muir moves to the Yosemite Valley.

Geologist and explorer John Wesley Powell travels the Colorado River through the Grand Canyon.

1872 Yellowstone National Park is established as the first national park of the United States in Yellowstone, Wyoming.

U.S. legislation: Passage of the Mining Law permits individuals to purchase rights to mine public lands.

1876 The Appalachian Mountain Club is founded.

1879 The U.S. Geological Survey (USGS) is formed.

1882 The first hydroelectric plant opens on the Fox River in Wisconsin.

1883 Krakatoa, a small island of Indonesia, is virtually destroyed by a volcanic explosion.

1890 Denmark constructs the first windmill for use in generating electricity.

Sequoia National Park, Yosemite National Park, and General Grant National Park are established in California.

1891 U.S. legislation: Passage of Forest Reserve Act provides the basis for a system of national forests.

1892 John Muir, Robert Underwood Johnson, and William Colby are cofounders of the Sierra Club, in Muir's words, to "do something for wildness and make the mountains glad."

1893 The National Trust is founded in the United Kingdom. The group purchases land deemed of having natural beauty or considered a cultural landmark.

1895 Founding of the American Scenic and Historic Preservation Society.

1898 Cornell University establishes the first college program in forestry.

Gifford Pinchot becomes head of the U.S. Division of Forestry (now the U.S. Forest Service) and serves until 1910. Under President

Theodore Roosevelt, many of Pinchot's ideas became national policy. During his service, the national forests increase from 32 in 1898 to 149 in 1910, a total of 193 million acres.

1899 The River and Harbor Act bans pollution of all navigable waterways. Under the act, the building of any wharves, piers, jetties, and other structures is prohibited without congressional approval.

1900 U.S. legislation: Passage of Lacey Act makes it unlawful to transport illegally killed game animals across state boundaries.

1902 U.S. legislation: Passage of Reclamation Act establishes the Bureau of Reclamation.

1903 First federal U.S. wildlife refuge is established on Pelican Island in Florida.

1905 The National Audubon Society, named for wildlife artist John James Audubon, is founded.

1906 Yosemite Valley is incorporated into Yosemite National Park.

1907 International Association for the Prevention of Smoke is founded. The group's name later changes several times to reflect other concerns over causes of air pollution.

Gifford Pinchot is appointed the first chief of the U.S. Forest Service.

1908 The Grand Canyon is set aside as a national monument.

Chlorination is first used at U.S. water treatment plants.

President Theodore Roosevelt hosts the first Governors' Conference on Conservation.

1914 The last passenger pigeon, Martha, dies in the Cincinnati zoo.

1916 The National Park Service (NPS) is established.

1918 Hunting of migratory bird species is restricted through passage of the Migratory Bird Treaty Act. The act supports treaties between the United States and surrounding nations.

Save-the-Redwoods League is created.

1920 U.S. legislation: Passage of the Mineral Leasing Act regulates mining on federal lands.

1922 The Izaak Walton League is organized under the direction of **Will H. Dilg**.

1924 Environmentalist **Aldo Leopold** wins designation of Gila National Forest, New Mexico, as first extensive wilderness area.

Marjory Stoneman Douglas, of the *Miami Herald*, writes newspaper columns opposing the draining of the Florida Everglades.

Bryce Canyon National Park is established in Utah.

1925 The Geneva Protocol is signed by numerous countries as a means of stopping use of biological weapons.

1928 The Boulder Canyon Project (Hoover Dam) is authorized to provide irrigation, electric power, and a flood-control system for Arizona and Nevada communities.

1930 Chlorofluorocarbons (CFCs) are deemed safe for use in refrigerators and air conditioners.

1931 France builds and makes use of the first Darrieus aerogenerator to produce electricity from wind energy.

Addo Elephant National Park is established in the Eastern Cape region of South Africa to provide a protected habitat for African elephants.

1932 Hugh Bennett is given the opportunity to put his soil conservation ideas into practice to help reduce soil erosion. He becomes the director of the Soil Erosion Service (SES) created by the Department of Interior.

1933 The Tennessee Valley Authority (TVA) is formed.

The Civilian Conservation Corps (CCC) employs more than 2 million Americans in forestry, flood control, soil erosion, and beautification projects.

1934 The greatest drought in U.S. history continues. Portions of Texas, Oklahoma, Arkansas, and several other midwestern states are known as the "Dust Bowl."

U.S. legislation: Passage of Taylor Grazing Act regulates livestock grazing on federal lands.

1935 The Soil Conservation Service (SCS) is established.

The Wilderness Society is founded.

1936 The National Wildlife Federation (NWF) is formed.

1939 David Brower produces his first nature film for the Sierra Club, called *Sky Land Trails of the Kings*. In the same year, Brower, who is an excellent climber, completes his most famous ascent, Shiprock, a volcanic plug which rises 1,400 feet from the floor of the New Mexico desert.

1940 The U.S. Wildlife Service is established to protect fish and wildlife.

U.S. legislation: President Franklin Roosevelt signs the Bald Eagle Protection Act.

1945 The United Nations (UN) establishes the Food and Agriculture Organization (FAO).

1946 The International Whaling Commission (IWC) is formed to research whale populations.

The U.S. Bureau of Land Management (BLM) and the Atomic Energy Commission (AEC) are created.

1947 Marjory Stoneman Douglas publishes *The Everglades: River of Grass* and serves as a member of the committee that gets the Everglades designated a national park.

1948 The UN creates the International Union for the Conservation of Nature (IUCN) as a special environmental agency.

An air pollution incident in Donora, Pennsylvania, kills 20 people; 14,000 become ill.

U.S. legislation: Passage of Federal Water Pollution Control Law.

1949 Aldo Leopold's *A Sand County Almanac* is published posthumously.

1950 Oceanographer **Jacques Cousteau** purchases and transforms a former minesweeper, the *Calypso*, into a research vessel which he uses to increase awareness of the ocean environment.

1951 Tanzania begins its national park system with the establishment of the Serengeti National Park.

1952 Clean air legislation is enacted in Great Britain after air pollution–induced smog brings about the deaths of nearly 4,000 people.

David Brower becomes the first executive director of the Sierra Club.

1953 Radioactive iodine from atomic bomb testing is found in the thyroid glands of children living in Utah.

1955 U.S. legislation: Passage of the Air Pollution Control Act, the first federal legislation designed to control air pollution.

1956 U.S. legislation: Passage of the Water Pollution Control Act authorizes development of water-treatment plants.

1959 The Antarctic Treaty is signed to preserve natural resources of the continent.

1961 The African Wildlife Foundation (AWF) is established as an international organization to protect African wildlife.

1962 Rachel Carson publishes *Silent Spring*, a groundbreaking study of the dangers of DDT and other insecticides.

Hazel Wolf joins the National Audubon Society in Seattle, Washington, and plays a prominent role in local, national, and international environmental efforts during her lifetime.

1963 The Nuclear Test Ban Treaty between the United States and the Soviet Union stops atmospheric testing of nuclear weapons.

U.S. legislation: Passage of the first Clean Air Act (CAA) authorizes money for air pollution control efforts.

1964 Hazel Henderson organizes women in a local play park in New York City and starts a group called Citizens for Clean Air, the first environmental group, she believes, east of the Mississippi. She built Citizens for Clean Air from a very small group to a membership of 40,000. Two years later, 80 people died in New York City from air pollution–related causes during four days of atmospheric inversion.

U.S. legislation: Passage of the Wilderness Act creates the National Wilderness Preservation System.

1965 U.S. legislation: Passage of the Water Quality Act authorizes the federal government to set water standards in absence of state action.

1966 Eighty people in New York City die from air pollution–related causes.

1967 The *Torey Canyon* runs aground spilling 175 tons of crude oil off Cornwall, England.

Dian Fossey establishes the Karisoke Research Center in the Virunga Mountains, within the Parc National des Volcans in Rwanda to study endangered mountain gorillas.

The Environmental Defense Fund (EDF) is formed to lead an effort to save the osprey from DDT.

1968 U.S. legislation: Passage of the Wild and Scenic Rivers Act and the National Trails System Act identify areas of great scenic beauty for preservation and recreation.

Paul Ehrlich publishes *The Population Bomb*.

1969 Wildlife photographer Joy Adamson establishes the Elsa Wild Animal Appeal, an organization dedicated to the preservation and humane treatment of wild and captive animals.

Greenpeace is created.

Blowout of oil well in Santa Barbara, California, releases 2,700 tons of crude oil into the Pacific Ocean.

U.S. legislation: Passage of the National Environmental Policy Act (NEPA) requires all federal agencies to complete an environmental impact statement for any dam, highway, or other large construction project undertaken, regulated, or funded by the federal government.

The Friends of the Earth (FOE) is founded in the United States.

John Todd, **Nancy Jack Todd**, and Bill McLarney are the cofounders of the New Alchemy Institute in Cape Cod, Massachusetts. The institute begins to pioneer a new way of treating sewage and other wastes.

1970 Denis Hayes is the national coordinator of the first Earth Day, which is celebrated on April 22.

Construction of the Aswan High Dam on the Nile River in Egypt is completed.

U.S. legislation: Passage of an amended Clean Air Act (CAA) expands air pollution control.

The U.S. Environmental Protection Agency (EPA) is established.

1971 Canadian primatologist Biruté Galdikas begins her studies of orangutans through the Orangutan Research and Conservation Project in Borneo.

The United Nations Educational, Scientific and Cultural Organization (UNESCO) establishes the Man and the Biosphere Program, developing a global network of biosphere reserves.

1972 The Biological and Toxin Weapons Convention is adopted by 140 nations to stop the use of biological weapons.

The EPA phases out the use of DDT in the United States to protect several species of predatory birds. The ban builds on information obtained from Rachel Carson's 1962 book, *Silent Spring*.

U.S. legislation: Passage of the Water Pollution Control Act, the Coastal Zone Management Act (CZMA), and the Environmental Pesticide Control Act.

Oregon passes the first bottle-recycling law.

1973 Norwegian philosopher Arne Naess coins the term *deep ecology* to describe his belief that humans need to recognize natural things for their intrinsic value, rather than just for their value to humans.

The Convention on International Trade in Endangered Species of Wild Fauna and Flora (CITES) is signed by more than 80 nations. The Endangered Species Act of the United States also is enacted.

Congress approves construction of the 1,300-kilometer pipeline from Alaska's North Slope oil field to the Port of Valdez.

An Energy crisis in the United States arises from an Arab oil embargo.

A collision and resulting explosion between the *Corinthos* oil tanker and the *Edgar M. Queeny* releases 272,000 barrels of crude oil and other chemicals into the Delaware River near Marcus Hook, Pennsylvania.

1974 Scientists report their discovery of a hole in the ozone layer above Antarctica.

U.S. legislation: Passage of the Safe Drinking Water Act sets standards to protect the nation's drinking water. The EPA bans most uses for aldrin and dieldrin and disallows the production and importation of these chemicals into the United States.

1975 Unleaded gas goes on sale. New cars are equipped with antipollution catalytic converters.

The EPA bans use of asbestos insulation in new buildings.

Edward Abbey publishes *The Monkey Wrench Gang*, a novel detailing acts of ecotage as a means of protecting the environment.

1976 *Argo Merchant* runs aground releasing 25,000 tons of fuel into the Atlantic Ocean near Nantucket, Rhode Island.

National Academy of Sciences reports that CFC gases from spray cans are damaging the ozone layer.

U.S. legislation: Passage of the Resource Conservation and Recovery Act empowers the EPA to regulate the disposal and treatment of municipal solid and hazardous wastes. The Toxic Substances Control Act and the Resource Conservation and Recovery Act are enacted.

Fire aboard the *Hawaiian Patriot* releases nearly 100,000 tons of crude oil into the Pacific Ocean.

1977 The Green Belt Movement is begun by Kenyan conservationist Wangari Muta Maathai on World Environment Day.

Blowout of Ekofisk oil well releases 27,000 tons of crude oil into the North Sea.

Construction of the Alaska pipeline, the 1,300-kilometer pipeline that carries oil from

Alaska's North Slope oil field to the Port of Valdez, is completed at a cost of more than $8 billion.

U.S. legislation: Passage of the Surface Mining Control and Reclamation Act.

The Department of Energy (DOE) is created.

1978 The *Amoco Cadiz* tanker runs aground spilling 226,000 tons of oil into the ocean near Portsall, Brittany.

People living in the Love Canal community of New York are evacuated from the area to reduce their exposure to chemical wastes which have surfaced from a canal formerly used as a dump site.

Rainfall in Wheeling, West Virginia, is measured at a pH of 2, the most acidic rain yet recorded.

Aerosols with fluorocarbons are banned in the United States.

The EPA bans the use of asbestos in insulation, fireproofing, or decorative materials.

1979 British scientist **James E. Lovelock** publishes *Gaia: A New Look at Life on Earth.*

Collision of the *Atlantic Empress* and the *Aegean Captain* releases 370,000 tons of oil into the Caribbean Sea.

The Convention on Long-Range Transboundary Air Pollution (LRTAP) is signed by several European nations to limit sulfur dioxide emissions which cause acid rain problems in other countries.

The Three Mile Island Nuclear Power Plant in Pennsylvania experiences near-meltdown.

The EPA begins a program to assist states in removing flaking asbestos insulation from pipes and ceilings in school buildings throughout the United States.

The EPA bans the marketing of herbicide Agent Orange in the United States.

1980 Debt-for-nature swap idea is proposed by Thomas E. Lovejoy: nations could convert debt to cash which would then be used to purchase parcels of tropical rain forest to be managed by local conservation groups.

Global Report to the President addresses world trends in population growth, natural resource use, and the environment by the end of the century, and calls for international cooperation in solving problems.

U.S. legislation: Passage of the Comprehensive Environmental Response, Compensation, and Liability Act (Superfund) and the Low Level Radioactive Waste Policy Act.

1981 Earth First!, a radical environmental action group that resorts to ecotage to gain its objectives, formed.

Lois Gibbs founds the Citizens' Clearinghouse for Hazardous Wastes, later named the Center for Health, Environment, and Justice (CHEJ).

1982 U.S. legislation: Passage of the Nuclear Waste Policy Act.

1983 A film of **Randy Hayes**, *The Four Corners, a National Sacrifice Area,* wins the 1983 Student Academy Award for the best documentary. The film documents the tragic effects of uranium and coal mining on Hopi and Navajo Indian lands in the American Southwest.

The residents of Times Beach, Missouri, are ordered to evacuate their community. Investigations of Times Beach in the 1980s disclosed the fact that oil contaminated with dioxin, a highly toxic substance, had been used to treat the town's streets.

Cathrine Sneed founds and acts as director of the Garden Project in San Francisco. The Garden Project, a horticulture class for inmates of the San Francisco County Jail, uses organic gardening as a metaphor for life change. The U.S. Department of Agriculture calls the project "one of the most innovative and successful community-based crime prevention programs in the country."

1984 Toxic gases released from the Union Carbide chemical manufacturing plant in Bhopal, India kill an estimated 3,000 people and injure thousands of others.

The Jane Goodall Institute (JGI) is founded.

The British tanker *Alvenus* spills 0.8 million gallons of oil into the Gulf of Mexico.

U.S. legislation: Passage of the Hazardous and Solid Waste Amendments.

1985 Concerned Citizens of South Central Los Angeles becomes one of the first African American environmental groups in the United States. **Julia Tate** serves as the executive director. The organization's goal is to provide a better quality of life for the residents of this Los Angeles community. **Maria Perez**, **Nevada Dove**, and **Fabiola Tostado** later join the group and are known as the Toxic Crusaders.

Huey D. Johnson becomes the founder and president of the Resource Renewal Institute

(RRI), a nonprofit organization located in California. Johnson suggests that green plans is the path countries should take to respond to environmental decline. Green plans treat the environment as it really exists—a single, interconnected ecosystem that can be safeguarded for future generations only through a systemic, long-range plan of action.

Scientists of the British Antarctica Survey discover the ozone hole. The hole, which appears during the Antarctic spring, is caused by the chlorine from CFCs.

Juana Gutiérrez becomes president and founder of Mothers of East Los Angeles, Santa Isabel Chapter (Madres del Este de Los Angeles—Santa Isabel) (MELASI) whose mission is to fight against toxic dumps and incinerators and also to take a proactive approach to community improvement.

Primatologist Dian Fossey is discovered murdered in her cabin at the Karosoke Research Center she founded. Her death is attributed to poachers.

While protesting nuclear testing being conducted by France in the Pacific Ocean, the *Rainbow Warrior* (a boat owned by Greenpeace) is sunk in a New Zealand harbor by agents of the French government.

U.S. legislation: Passage of the Food Security Act.

1986 Tons of toxic chemicals stored in a warehouse owned by the Sandoz pharmaceutical company are released into the Rhine River near Basel, Switzerland. The effects of the spill are experienced in Switzerland, France, Germany, and Luxembourg.

An explosion destroys a nuclear power plant in Chernobyl, Ukraine, immediately killing more than 30 people and leading to the permanent evacuations of more than 100,000 others.

Bovine spongiform encephalopathy (BSE), a neurodegenerative illness of cattle, also known as mad cow disease, comes to the attention of the scientific community when it appears in cattle in the United Kingdom.

U.S. legislation: Passage of the Emergency Response and Community Right-to-Know Act and the Superfund Amendments and Reauthorization Act (SARA).

1987 The Montreal Protocol, an international treaty that proposes to cut in half the production and use of CFCs, is approved by more than 30 nations.

The world's fourth largest lake, the Aral Sea of Asia, is divided in two as a result of the diversion of water from its feeder streams, the Syr Darya and Amu Darya rivers.

The *Mobro*, a garbage barge from Long Island, New York, travels 9,600 kilometers in search of a place to offload the garbage it carries.

1988 Use of ruminant proteins in the preparation of cattle feed is banned in the United Kingdom to prevent outbreaks of BSE.

Global temperatures reach their highest levels in 130 years.

The Ocean Dumping Ban legislates international dumping of wastes in the ocean.

EPA studies report that indoor air can be 100 times as polluted as outdoor air. Radon is found to be widespread in U.S. homes.

Beaches on the east coast of the United States are closed because of contamination by medical waste washed onshore.

The United States experiences its worst drought in 50 years.

Plastic ring six-pack holders are required to be made degradable.

U.S. legislation: Passage of the Plastic Pollution Research and Control Act bans ocean dumping of plastic materials.

1989 The United Kingdom bans the use of cattle brains, spinal cords, tonsils, thymuses, spleens, and intestines in foods intended for human consumption as a means of preventing further outbreaks of Creutzfeldt-Jakob disease (CJD), the human version of mad cow disease, in humans.

Fire aboard the *Kharg 5* releases 75,000 tons of oil into the sea surrounding the Canary Islands.

The Montreal Protocol treaty is updated and amended.

The New York Department of Environmental Conservation reports that 25 percent of the lakes and ponds in the Adirondacks are too acidic to support fish.

The *Exxon Valdez* runs aground on Prince William Sound, Alaska, spilling 11 million gallons of oil into one of the world's most fragile ecosystems.

1990 Ocean Robbins, age 16, and **Ryan Eliason**, 18, are the cofounders of YES!, or Youth for Environmental Sanity. The goal of YES! is to educate, inspire, and empower young people to take positive action for healthy people and a healthy planet. Robbins served as director for five years and is now the organization's president. As of 2000, the program has reached 600,000 students in 1,200 schools in 43 states through full school assemblies.

UN report forecasts a world temperature increase of 2°F within 35 years as a result of greenhouse gas emissions.

U.S. legislation: Passage of the Clean Air Act amendments including requirements to control the emission of sulfur dioxide and nitrogen oxides.

1991 The Gulf War concludes with hundreds of oil wells in Kuwait being set afire by Iraqi troops, resulting in extensive air and water pollution problems.

The United States accepts an agreement on Antarctica which prohibits activities relating to mining, protects native species of flora and fauna, and limits tourism and marine pollution.

Eight scientists begin a two-year stay in Biosphere 2 in Arizona, a test center designed to provide a self-sustaining habitat modeling Earth's natural environments. The experiment, which is repeated in 1993, meets with much criticism and is deemed largely unsuccessful.

1992 UN Earth Summit is held in Rio de Janeiro, Brazil. Major resolutions resulting from the summit include the Rio Declaration on Environment and Development, Agenda 21, Biodiversity Convention, Statement of Forest Principles, and the Global Warming Convention, which is signed by more than 160 nations.

Severn Cullis-Suzuki speaks for six minutes to the delegates urging them to work hard on resolving global environmental issues. She received a standing ovation.

The Montreal Protocol is again amended with signatories agreeing to phase out CFC use by the year 2000.

1993 Sugar producers and U.S. government agree on a restoration plan for the Florida Everglades.

1994 *Dumping in Dixie: Class and Environmental Quality* is published by **Robert Bullard**. The book reports on five environmental justice campaigns in states ranging from Texas to West Virginia. Bullard emphasizes that African Americans are concerned about and do participate in environmental issues.

The California Desert Protection Act is passed.

Failure of a dike results in the release of 102,000 tons of oil into the Siberian tundra near Usink in northern Russia.

The Russian government calls for preventive measures to control the destruction of Lake Baikail.

The bald eagle is reclassified from an endangered species to a threatened species on the U.S. Endangered Species List.

An 8.5-million-gallon spill is discovered in Unocal's Guadalupe oil field in California.

1995 The U.S. Government reintroduces endangered wolves to Yellowstone Park.

1999 Scientists report that the human population of Earth now exceeds 6 billion people.

The peregrine falcon is removed from the U.S. Endangered Species List.

The *New Carissa* runs aground off the coast of Oregon, leaking some oil into Coos Bay. The tanker is later towed into the open ocean and sunk.

Beyond Globalization: Shaping a Sustainable Global Economy is published by Hazel Henderson.

Paul Hawken coauthors *Natural Capitalism, Creating the Next Industrial Revolution.*

Off the Map, an Expedition Deep into Imperialism, the Global Economy, and Other Earthly Whereabouts is published by **Chellis Glendinning**.

Twenty-three-year-old **Julia Butterfly Hill** comes down out of a 180-foot California redwood tree after living there for two years to prevent the destruction of the forest. A deal is made with the logging company to spare the tree as well as a three-acre buffer zone.

2000 Denis Hayes is the coordinator and **Mark Dubois** is the international coordinator of Earthday 2000.

Ralph Nader and **Winona LaDuke** run for U.S. president and vice president on the Green Party ticket.

In January 2000, **Hazel Wolf** passes away at the age of 101.

The Chernobyl nuclear power plant is scheduled to close in December.

Anthropologists for the Wildlife Conservation Society in New York announce that a type of large West African monkey is extinct, making it the first primate to vanish in the twenty-first century.

A study by National Park Trust, a privately funded land conservancy, finds that more than 90,000 acres within state parks in 32 states are threatened by commercial and residential development and increased traffic, among other things.

A bone-dry summer in north-central Texas breaks the Depression-era drought record when the Dallas area marks 59 days without rain. The arid streak with 100-degree daily highs breaks a record of 58 days set in the midst of the Dust Bowl in 1934 and tied in 1950. The Texas drought exceeded 1 billion dollars in agricultural losses.

Massachusetts announces that the state will spend $600,000 to determine whether petroleum pollution in largely African American city neighborhoods contributes to lupus, a potentially deadly immune disease. The research, to be conducted over three years, will target three areas of the city with unusually high levels of petroleum contamination.

Hybrid vehicle Toyota Prius is offered for sale in the United States.

The hole in the ozone layer over Antarctica has stretched over a populated city for the first time, after ballooning to a new record size. Previously, the hole had opened only over Antarctica and the surrounding ocean.

2001 An environmental group that successfully campaigned for the return of wolves to Yellowstone National Park wants the federal government to do the same in western Colorado and parts of Utah, southern Wyoming, northern New Mexico, and Arizona.

The UN Environment Program launches a campaign to save the world's great apes from extinction, asking for at least $1 million to get started.

The captain and crew of a tanker that spilled at least 185,000 gallons of diesel into the fragile marine environment of the Galapagos Islands have been arrested.

One hundred sixty-five countries approve the Kyoto rules aimed at halting global warming. The Kyoto Protocol requires industrial countries to scale back emissions of carbon dioxide and other greenhouse gases by an average of 5 percent from their 1990 levels by 2012. The United States, the world's biggest polluter rejects the pact.

The EPA reaches an agreement for the phaseout of a widely used pesticide, diazinon, because of potential health risks to children.

For the second time in three years, the average fuel economy of new passenger cars and light trucks sold in the United States dropped to its lowest level since 1980.

More and more Americans are breathing dirtier air, and larger U.S. cities such as Los Angeles and Atlanta remain among the worst for pollution.

In rural stretches of Alaska, global warming has thinned the Arctic pack ice, making travel dangerous for native hunters. Traces of industrial pollution from distant continents is showing up in the fat of Alaska's marine wildlife and in the breast milk of native mothers who eat a traditional diet including seal and walrus meat.

2002 A Congo volcano devastates a Congolese town burning everything in its path, creating a five-foot-high wall of cooling stone, and leaving a half million people homeless.

New research is conducted in the practice of killing sharks solely for their fins.

A report by the USGS shows the nation's waterways are awash in traces of chemicals used in beauty aids, medications, cleaners, and foods. Among the substances are caffeine, painkillers, insect repellent, perfumes, and nicotine. These substances largely escape regulation and defy municipal wastewater treatment.

A microbe is discovered to be a major cause of the destruction of beech trees in the northeastern United States.

A study discovers that, if fallen leaves are left in stagnant water, they can release toxic mercury, which eventually can accumulate in fish that live far downstream.

Scientists are experimenting with various sprays containing clay particles to kill toxic algae in seawater.

Meteorologists discover that the Mediterranean Sea receives air current pollutants from Europe, Asia, and North America.

Researchers report possible ways of blocking the deadly effects of anthrax.

2003 A new international treaty—The Protocol on Persistent Organic Pollutants (POPS) was ratified by 17 nations although the United States has not signed on. The treaty drafted by United Nations reduces and eliminates 16 toxic chemicals that are long-lived in the environment and travel globally. The new treaty, an extension of an earlier one signed in 2000, added four more organic persistent pollutants to the list.

Many global scientific studies reveal that excessive ultraviolet (UV) sunlight and pollution are linked to a decline in amphibian populations. Now Canadian biologists find that too much exposure of excessive UV radiation to tadpole populations reduces their chances of becoming frogs.

2003 marked the 50th anniversary of the research and publication of a different structure of the DNA model proposed by James D. Watson and

Francis H.C. Crick. In 1953 the scientists reported that the DNA molecule resembled a spiral staircase.

A new excavation in South Africa discovered the oldest fossils in the human family. The bones of a skull and a partial arm found in two caves date back to 4 million years ago according to scientists in Johannesburg

Scientists in New Jersey discovered that some outdoor antimosquito coils used to keep insects away can also cause respiratory health problems. The spiral-shaped container releases pollutants in the fumes expelled from coil. The researchers suggest that consumers should check these products carefully.

Researchers in Australia reported that pieces of plastic litter found in oceans continue to have an effect on marine wildlife. Small plastic chips are a hazard for seabirds who mistake the litter for food or fish eggs. The litter also moves up the food chain from fish that have ingested the plastic chips and in turn seals eat them.

2004 A scientific study reported that consumers should limit their consumption of farm-raised Atlantic salmon because of high concentrations of chlorinated organic contaminants in the fish. Their study revealed that the farm-raised salmon were contaminated with polychlorinated biphenyls (PCBs) and other organic chemicals. Except for the PCBs, the researchers agree that the farm-raised fish are healthy but consumption should be limited to no more than once a month in the diet. The researchers based their dietary report on the U.S. Environmental Protection Agency cancer risk assessments.

A group in Salisbury Plain, England is restoring Stonehenge to its natural setting. As a popular historic site to visitors, Stonehenge had become an area surrounded by roads and parking lots. The new restoration plan calls for building an underground tunnel for traffic and removing one of the roads. The present parking lots will become open grassy lawns.

Experts reported that two billion people lack reliable access to safe and nutritious food and 800 million, 40 percent of them children, are classified as chronically malnourished.

Public health officials in Uganda have reported progress in the country's fight against HIV, the AIDS virus. Since 1990's HIV cases in Uganda have dropped by more than 60 percent. Unfortunately, Uganda's neighboring countries are not doing well in their HIV prevention programs.

United Nations Secretary-General Kofi Annan stated, "by 2025, two-thirds of the world's population may be living in countries that face serious water shortages." The growing population is making surface water scarcer particularly in urban areas.

United States and Israel scientists have found a way to produce hydrogen from water. The hydrogen energy can be used in making fuel cells to power vehicles and homes. The research team uses solar radiation to heat sodium hydroxide in a solution of water. At high temperatures the water molecules (H_2O) break apart into oxygen and hydrogen. Using solar-power to produce hydrogen is better environmentally than hydrogen derived from fossil fuels.

Appendix B: Endangered Species by State

The list below, obtained from the U.S. Fish and Wildlife Service, is an abridged listing of a selected group of endangered species (E) for each state. For a full list of endangered and threatened species, and other information about endangered species and the Endangered Species Act, see the Endangered Species Program Website at http://endangered.fws.gov/

ALABAMA

(Alabama has 106 plant and animal species that are listed as endangered (E) or threatened (T). The following list is only a selection of those plants and animals that are endangered. Contact the U.S. Fish and Wildlife Service to see the entire list.)

Animals

E - Bat, gray
E - Bat, Indiana
E - Cavefish, Alabama
T - Chub, spotfin
E - Clubshell, black
E - Combshell, southern
E - Darter, boulder
E - Fanshell
E - Kidneyshell, triangular
E - Lampmussel, Alabama
E - Manatee, West Indian
E - Moccasinshell, Coosa
E - Mouse, Alabama beach
E - Mussel, ring pink
E - Pearlymussel, cracking
E - Pearlymussel, Cumberland monkeyface
E - Pigtoe, dark
E - Plover, piping
E - Shrimp, Alabama cave
E - Snail, tulotoma (Alabama live-bearing)
E - Stork, wood
E - Turtle, Alabama redbelly (red-bellied)
E - Turtle, leatherback sea
E - Woodpecker, red-cockaded

Plants

E - Grass, Tennessee yellow-eyed
E - Leather-flower, Alabama
E - Morefield's leather-flower
E - Pinkroot, gentian
E - Pitcher-plant, Alabama canebrake
E - Pitcher-plant, green
E - Pondberry
E - Prairie-clover, leafy

ALASKA
Animals

E - Curlew, Eskimo (*Numenius borealis*)
E - Falcon, American peregrine (*Falco peregrinus anatum*)

Plant

E - Aleutian shield-fern (Aleutian holly-fern) (*Polystichum aleuticum*)

ARIZONA
Animals

E - Ambersnail, Kanab
E - Bat, lesser (Sanborn's) long-nosed
E - Bobwhite, masked (quail)
E - Chub, bonytail
E - Chub, humpback
E - Chub, Virgin River
E - Chub, Yaqui
E - Flycatcher, Southwestern willow
E - Jaguarundi
E - Ocelot
E - Pronghorn, Sonoran
E - Pupfish, desert
E - Rail, Yuma clapper
E - Squawfish, Colorado
E - Squirrel, Mount Graham red
E - Sucker, razorback
E - Topminnow, Gila (incl. Yaqui)
E - Trout, Gila
E - Vole, Hualapai Mexican
E - Woundfin

Plants

E – Arizona agave
E – Arizona cliffrose
E – Arizona hedgehog cactus
E – Brady pincushion cactus
E – Kearney's blue-star
E – Nichol's Turk's head cactus
E – Peebles Navajo cactus
E – Pima pineapple cactus
E – Sentry milk-vetch

ARKANSAS

Animals

E – Bat, gray
E – Bat, Indiana
E – Bat, Ozark big-eared
E – Beetle, American burying (giant carrion)
E – Crayfish, cave
E – Pearlymussel, Curtis'
E – Pearlymussel, pink mucket
E – Pocketbook, fat
E – Pocketbook, speckled
E – Rock-pocketbook, Ouachita (Wheeler's pearly mussel)
E – Sturgeon, pallid
E – Tern, least
E – Woodpecker, red-cockaded

Plants

E – Harperella
E – Pondberry
E – Running buffalo clover

CALIFORNIA

(California has more than 160 plant and animal species that are listed as endangered or threatened. The following list is only a selection of those plants and animals that are endangered. Contact the U.S. Fish and Wildlife Service to see the entire list.)

Animals

E – Butterfly, El Segundo blue
E – Butterfly, Lange's metalmark
E – Chub, Mohave tui
E – Condor, California
E – Crayfish, Shasta (placid)
E – Fairy shrimp, Conservancy
E – Falcon, American peregrine
E – Fly, Delhi Sands flower-loving
E – Flycatcher, Southwestern willow
E – Fox, San Joaquin kit
E – Goby, tidewater
E – Kangaroo rat, Fresno
E – Lizard, blunt-nosed leopard
E – Mountain beaver, Point Arena
E – Mouse, Pacific pocket
E – Pelican, brown
E – Pupfish, Owens
E – Rail, California clapper
E – Salamander, Santa Cruz long-toed
E – Shrike, San Clemente loggerhead
E – Shrimp, California freshwater
E – Snail, Morro shoulderband (banded dune)
E – Snake, San Francisco garter
E – Stickleback, unarmored threespine
E – Sucker, Lost River
E – Tadpole shrimp, vernal pool
E – Tern, California least
E – Toad, arroyo southwestern
E – Turtle, leatherback sea
E – Vireo, least Bell's
E – Vole, Amargosa

Plants

E – Antioch Dunes evening-primrose
E – Bakersfield cactus
E – Ben Lomond wallflower
E – Burke's goldfields
E – California jewelflower
E – California Orcutt grass
E – Clover lupine
E – Cushenbury buckwheat
E – Fountain thistle
E – Gambel's watercress
E – Kern mallow
E – Loch Lomond coyote-thistle
E – Robust spineflower (includes Scotts Valley spineflower)
E – San Clemente Island larkspur
E – San Diego button-celery
E – San Mateo thornmint
E – Santa Ana River woolly-star
E – Santa Barbara Island liveforever
E – Santa Cruz cypress
E – Solano grass
E – Sonoma sunshine (Baker's stickyseed)

E – Stebbins' morning-glory
E – Truckee barberry
E – Western lily

COLORADO

Animals

E – Butterfly, Uncompahgre fritillary
E – Chub, bonytail
E – Chub, humpback
E – Crane, whooping
E – Ferret, black-footed
E – Flycatcher, Southwestern willow
E – Plover, piping
E – Squawfish, Colorado
E – Sucker, razorback
E – Tern, least
E – Wolf, gray

Plants

E – Clay-loving wild-buckwheat
E – Knowlton cactus
E – Mancos milk-vetch
E – North Park phacelia
E – Osterhout milk-vetch
E – Penland beardtongue

CONNECTICUT

Animals

E – Mussel, dwarf wedge
E – Plover, piping
E – Tern, roseate
E – Turtle, hawksbill sea
E – Turtle, Kemp's (Atlantic) ridley sea
E – Turtle, leatherback sea

Plant

E – Sandplain gerardia

DELAWARE

Animals

E – Plover, piping
E – Squirrel, Delmarva Peninsula fox
E – Turtle, hawksbill sea
E – Turtle, Kemp's (Atlantic) ridley sea—Turtle,
   green sea
E – Turtle, leatherback sea

Plant

E – Canby's dropwort

FLORIDA

(Florida has more than 90 plant and animal species that are listed as endangered or threatened. The following list is only a selection of those plants and animals that are endangered. Contact the U.S. Fish and Wildlife Service to see the entire list.)

Animals

E – Bat, gray
E – Butterfly, Schaus swallowtail
E – Crocodile, American
E – Darter, Okaloosa
E – Deer, key
E – Kite, Everglade snail
E – Manatee, West Indian (Florida)
E – Mouse, Anastasia Island beach
E – Mouse, Choctawahatchee beach
E – Panther, Florida
E – Plover, piping
E – Rabbit, Lower Keys
E – Rice rat (silver rice rat)
E – Sparrow, Cape Sable seaside
E – Stork, wood
E – Tern, roseate
E – Turtle, hawksbill sea
E – Turtle, Kemp's (Atlantic) ridley sea
E – Turtle, leatherback sea
E – Vole, Florida salt marsh
E – Woodpecker, red-cockaded
E – Woodrat, Key Largo

Plants

E – Apalachicola rosemary
E – Beautiful pawpaw
E – Brooksville (Robins') bellflower
E – Carter's mustard
E – Chapman rhododendron
E – Cooley's water-willow
E – Crenulate lead-plant
E – Etonia rosemary
E – Florida golden aster
E – Fragrant prickly-apple
E – Garrett's mint
E – Key tree-cactus
E – Lakela's mint
E – Okeechobee gourd

E – Scrub blazingstar

E – Small's milkpea

E – Snakeroot

E – Wireweed

GEORGIA

Animals

E – Acornshell, southern

E – Bat, gray

E – Bat, Indiana

E – Clubshell, ovate

E – Clubshell, southern

E – Combshell, upland

E – Darter, amber

E – Darter, Etowah

E – Kidneyshell, triangular

E – Logperch, Conasauga

E – Manatee, West Indian (Florida)

E – Moccasinshell, Coosa

E – Pigtoe, southern

E – Plover, piping

E – Stork, wood

E – Turtle, hawksbill sea

E – Turtle, Kemp's (Atlantic) ridley sea

E – Turtle, leatherback sea

E – Woodpecker, red-cockaded

Plants

E – American chaffseed

E – Black-spored quillwort

E – Canby's dropwort

E – Florida torreya

E – Fringed campion

E – Green pitcher-plant

E – Hairy rattleweed

E – Harperella

E – Large-flowered skullcap

E – Mat-forming quillwort

E – Michaux's sumac

E – Persistent trillium

E – Pondberry

E – Relict trillium

E – Smooth coneflower

E – Tennessee yellow-eyed grass

HAWAII

(Hawaii has 300 plant and animal species listed as endangered or threatened. The following list is only a selection of those plants and animals that are endangered. Contact the U.S. Fish and Wildlife Service to see the entire list.)

Animals

E – 'Akepa, Hawaii (honeycreeper)

E – Bat, Hawaiian hoary

E – Coot, Hawaiian

E – Creeper, Hawaiian

E – Crow, Hawaiian

E – Duck, Hawaiian

E – Duck, Laysan

E – Finch, Laysan (honeycreeper)

E – Finch, Nihoa (honeycreeper)

E – Goose, Hawaiian (nene)

E – Hawk, Hawaiian

E – Millerbird, Nihoa (old world warbler)

E – Nukupu'u (honeycreeper)

E – Palila (honeycreeper)

E – Parrotbill, Maui (honeycreeper)

E – Petrel, Hawaiian dark-rumped

E – Snails, Oahu tree

E – Stilt, Hawaiian

E – Turtle, hawksbill sea

E – Turtle, leatherback sea

Plants

E – Abutilon eremitopetalum

E – Bonamia menziesii

E – Carter's panicgrass

E – Diamond Head schiedea

E – Dwarf iliau

E – Fosberg's love grass

E – Hawaiian bluegrass

E – Hawaiian red-flowered geranium

E – Kaulu

E – Kiponapona

E – Mahoe

E – Mapele

E – Nanu

E – Nehe

E – Opuhe

E – Pamakani

E – Round-leaved chaff-flower

E – Viola helenae

IDAHO

Animals

E – Caribou, woodland

E – Crane, whooping

E – Limpet, Banbury Springs
E – Snail, Snake River physa
E – Snail, Utah valvata
E – Springsnail, Bruneau Hot
E – Springsnail, Idaho
E – Sturgeon, white
E – Wolf, gray

Plants

(No plants on the endangered list)

ILLINOIS

Animals

E – Bat, gray
E – Bat, Indiana
E – Butterfly, Karner blue
E – Dragonfly, Hine's emerald
E – Falcon, American peregrine
E – Fanshell
E – Pearlymussel, Higgins' eye
E – Pearlymussel, orange-foot pimple back
E – Pearlymussel, pink mucket
E, T – Plover, piping
E – Pocketbook, fat
E – Snail, Iowa Pleistocene
E – Sturgeon, pallid
E – Tern, least

Plant

E – Leafy prairie-clover

INDIANA

Animals

E – Bat, gray
E – Bat, Indiana
E – Butterfly, Karner blue
E – Butterfly, Mitchell's satyr
E – Clubshell
E – Fanshell
E – Mussel, ring pink (golf stick pearly)
E – Pearlymussel, cracking
E – Pearlymussel, orange-foot pimple back
E – Pearlymussel, pink mucket
E – Pearlymussel, tubercled-blossom
E – Pearlymussel, white cat's paw
E – Pearlymussel, white wartyback
E – Pigtoe, rough
E, T – Plover, piping

E – Pocketbook, fat
E – Riffleshell, northern
E – Tern, least

Plant

E – Running buffalo clover

IOWA

Animals

E – Bat, Indiana
E – Pearlymussel, Higgins' eye
E – Plover, piping
E – Snail, Iowa Pleistocene
E – Sturgeon, pallid
E – Tern, least

Plants

(No plants on the endangered list)

KANSAS

Animals

E – Bat, gray
E – Bat, Indiana
E – Crane, whooping
E – Curlew, Eskimo
E – Ferret, black-footed
E – Plover, piping
E – Sturgeon, pallid
E – Tern, least
E – Vireo, black-capped

Plants

(No plants on endangered list)

KENTUCKY

Animals

E – Bat, gray
E – Bat, Indiana
E – Bat, Virginia big-eared
E – Clubshell
E – Darter, relict
E – Falcon, American peregrine
E – Fanshell
E – Mussel, ring pink (golf stick pearly)
E – Mussel, winged mapleleaf
E – Pearlymussel, cracking
E – Pearlymussel, Cumberland bean

E - Pearlymussel, dromedary
E - Pearlymussel, little-wing
E - Pearlymussel, orange-foot pimple back
E - Pearlymussel, pink mucket
E - Pearlymussel, purple cat's paw
E - Pearlymussel, tubercled-blossom
E - Pearlymussel, white wartyback
E - Pigtoe, rough
E - Plover, piping
E - Pocketbook, fat
E - Riffleshell, northern
E - Riffleshell, tan
E - Shiner, Palezone
E - Shrimp, Kentucky cave
E - Sturgeon, pallid
E - Tern, least
E - Woodpecker, red-cockaded

Plants

E - Cumberland sandwort
E - Rock cress
E - Running buffalo clover
E - Short's goldenrod

LOUISIANA

Animals

E - Manatee, West Indian (Florida)
E - Pearlymussel, pink mucket
E - Pelican, brown
E - Plover, piping
E - Sturgeon, pallid
E - Tern, least
T - Turtle, green sea
E - Turtle, hawksbill sea
E - Turtle, Kemp's (Atlantic) ridley sea
E - Turtle, leatherback sea
E - Vireo, black-capped
E - Woodpecker, red-cockaded

Plants

E - American chaffseed
E - Louisiana quillwort
E - Pondberry

MAINE

Animals

E - Plover, piping
E - Tern, roseate
E - Turtle, leatherback sea

Plant

E - Furbish lousewort

MARYLAND

Animals

E - Bat, Indiana
E - Darter, Maryland
E - Mussel, dwarf wedge
E - Plover, piping
E - Squirrel, Delmarva Peninsula fox
E - Turtle, hawksbill sea
E - Turtle, Kemp's (Atlantic) ridley sea
E - Turtle, leatherback sea

Plants

E - Canby's dropwort
E - Harperella
E - Northeastern (Barbed bristle) bulrush
E - Sandplain gerardia

MASSACHUSETTS

Animals

E - Beetle, American burying (giant carrion)
E - Falcon, American peregrine
E - Mussel, dwarf wedge
E - Plover, piping
E - Tern, roseate
E - Turtle, hawksbill sea
E - Turtle, Kemp's (Atlantic) ridley sea
E - Turtle, leatherback sea
E - Turtle, Plymouth redbelly (red-bellied)

Plants

E - Northeastern (Barbed bristle)
E - Sandplain gerardia

MICHIGAN

Animals

E - Bat, Indiana
E - Beetle, American burying (giant carrion)
E - Beetle, Hungerford's crawling water
E - Butterfly, Karner blue
E - Butterfly, Mitchell's satyr
E - Clubshell
E - Plover, piping
E - Riffleshell, northern
E - Warbler, Kirtland's
E - Wolf, gray

Plant

E – Michigan monkey-flower

MINNESOTA

Animals

E – Butterfly, Karner blue
E – Mussel, winged mapleleaf
E – Pearlymussel, Higgins' eye
E – Plover, piping
E – Wolf, gray

Plant

E – Minnesota trout lily

MISSISSIPPI

Animals

E – Bat, Indiana
E – Clubshell, black (Curtus' mussel)
E – Clubshell, ovate
E – Clubshell, southern
E – Combshell, southern (penitent mussel)
E – Crane, Mississippi sandhill
E – Falcon, American peregrine
E – Manatee, West Indian (Florida)
E – Pelican, brown
E – Pigtoe, flat (Marshall's mussel)
E – Pigtoe, heavy (Judge Tait's mussel)
E – Plover, piping
E – Pocketbook, fat
E – Stirrupshell
E – Sturgeon, pallid
E – Tern, least
E – Turtle, hawksbill sea
E – Turtle, Kemp's (Atlantic) ridley sea
E – Turtle, leatherback sea
E – Woodpecker, red-cockaded

Plants

E – American chaffseed
E – Pondberry

MISSOURI

Animals

E – Bat, gray
E – Bat, Indiana
E – Bat, Ozark big-eared
E – Pearlymussel, Curtis'

E – Pearlymussel, Higgins' eye
E – Pearlymussel, pink mucket
E – Plover, piping
E – Pocketbook, fat
E – Sturgeon, pallid
E – Tern, least

Plants

E – Missouri bladderpod
E – Pondberry
E – Running buffalo clover

MONTANA

Animals

E – Crane, whooping
E – Curlew, Eskimo
E – Ferret, black-footed
E – Plover, piping
E – Sturgeon, pallid
E – Sturgeon, white
E – Tern, least
E – Wolf, gray

Plants

(No plants on endangered list)

NEBRASKA

Animals

E – Beetle, American burying
   (giant carrion)
E – Crane, whooping
E – Curlew, Eskimo
E – Ferret, black-footed
E – Plover, piping
E – Sturgeon, pallid
E – Tern, least

Plant

E – Blowout penstemon

NEVADA

Animals

E – Chub, bonytail
E – Chub, Pahranagat roundtail (bonytail)
E – Chub, Virgin River
E – Cui-ui
E – Dace, Ash Meadows speckled

E – Dace, Clover Valley speckled
E – Dace, Independence Valley speckled
E – Dace, Moapa
E – Poolfish (killifish), Pahrump
E – Pupfish, Ash Meadows Amargosa
E – Pupfish, Devils Hole
E – Pupfish, Warm Springs
E – Spinedace, White River
E – Springfish, Hiko White River
E – Springfish, White River
E – Sucker, razorback
E – Woundfin

Plants

E – Amargosa niterwort
E – Steamboat buckwheat

NEW HAMPSHIRE

Animals

E – Butterfly, Karner blue
E – Mussel, dwarf wedge
E – Turtle, leatherback sea

Plants

E – Jesup's milk-vetch
E – Northeastern (Barbed bristle) bulrush
E – Robbins' cinquefoil

NEW JERSEY

Animals

E – Bat, Indiana
E – Plover, piping
E – Tern, roseate
E – Turtle, hawksbill sea
E – Turtle, Kemp's (Atlantic) ridley sea
E – Turtle, leatherback sea

Plant

E – American chaffseed

NEW MEXICO

Animals

E – Bat, lesser (Sanborn's) long-nosed
E – Bat, Mexican long-nosed
E – Crane, whooping
E – Gambusia, Pecos

E – Isopod, Socorro
E – Minnow, Rio Grande silvery
E – Springsnail, Alamosa
E – Springsnail, Socorro
E – Sucker, razorback
E – Tern, least
E – Topminnow, Gila (incl. Yaqui)
E – Trout, Gila
E – Woundfin

Plants

E – Holy Ghost ipomopsis
E – Knowlton cactus
E – Kuenzler hedgehog cactus
E – Lloyd's hedgehog cactus
E – Mancos milk-vetch
E – Sacramento prickly-poppy
E – Sneed pincushion cactus
E – Todsen's pennyroyal

NEW YORK

Animals

E – Butterfly, Karner blue
E – Mussel, dwarf wedge
E, T – Plover, piping
E – Tern, roseate
E – Turtle, hawksbill sea
E – Turtle, Kemp's (Atlantic) ridley sea
E – Turtle, leatherback sea

Plants

E – Northeastern (Barbed bristle) bulrush
E – Sandplain gerardia

NORTH CAROLINA

Animals

E – Bat, Indiana
E – Bat, Virginia big-eared
E – Butterfly, Saint Francis' satyr
E – Elktoe, Appalachian
E – Falcon, American peregrine
E – Heelsplitter, Carolina
E – Manatee, West Indian (Florida)
E – Mussel, dwarf wedge
E – Pearlymussel, little-wing
E – Plover, piping
E – Shiner, Cape Fear
E – Spider, spruce-fir moss

E – Spinymussel, Tar River
E – Squirrel, Carolina northern flying
E – Tern, roseate
E – Turtle, hawksbill sea
E – Turtle, Kemp's (Atlantic) ridley sea
E – Turtle, leatherback sea
E – Wolf, red
E – Woodpecker, red-cockaded

Plants

E – American chaffseed
E – Bunched arrowhead
E – Canby's dropwort
E – Cooley's meadowrue
E – Green pitcher-plant
E – Harperella
E – Michaux's sumac
E – Mountain sweet pitcher-plant
E – Pondberry
E – Roan Mountain bluet
E – Rock gnome lichen
E – Rough-leaved loosestrife
E – Schweinitz's sunflower
E – Small-anthered bittercress
E – Smooth coneflower
E – Spreading avens
E – White irisette

NORTH DAKOTA

Animals

E – Crane, whooping
E – Curlew, Eskimo
E – Falcon, American peregrine
E – Ferret, black-footed
E – Plover, piping
E – Sturgeon, pallid
E – Tern, least
E – Wolf, gray

Plants

(No plants on endangered list)

OHIO

Animals

E – Bat, Indiana
E – Beetle, American burying (giant carrion)
E – Butterfly, Karner blue
E – Butterfly, Mitchell's satyr

E – Clubshell
E – Dragonfly, Hine's emerald
E – Fanshell
E – Madtom, Scioto
E – Pearlymussel, pink mucket
E – Pearlymussel, purple cat's paw
E – Pearlymussel, white cat's paw
E, T – Plover, piping
E – Riffleshell, northern

Plant

E – Running buffalo clover

OKLAHOMA

Animals

E – Bat, gray
E – Bat, Indiana
E – Bat, Ozark big-eared
E – Beetle, American burying
    (giant carrion)
E – Crane, whooping
E – Curlew, Eskimo
E – Plover, piping
E – Rock-pocketbook, Ouachita
E – Tern, least
E – Vireo, black-capped
E – Woodpecker, red-cockaded

Plants

(No plants on the endangered list)

OREGON

Animals

E – Chub, Borax Lake
E – Chub, Oregon
E – Deer, Columbian white-tailed
E – Pelican, brown
E – Sucker, Lost River
E – Sucker, shortnose
E – Turtle, leatherback sea

Plants

E – Applegate's milk-vetch
E – Bradshaw's desert-parsley
E – Malheur wire-lettuce
E – Marsh sandwort
E – Western lily

PENNSYLVANIA

Animals

E – Bat, Indiana
E – Clubshell
E – Mussel, dwarf wedge
E – Mussel, ring pink (golf stick pearly)
E – Pearlymussel, cracking
E – Pearlymussel, orange-foot pimple back
E – Pearlymussel, pink mucket
E – Pigtoe, rough
E,T – Plover, piping
E – Riffleshell, northern

Plant

E – Northeastern (Barbed bristle)
    bulrush

RHODE ISLAND

Animals

E – Beetle, American burying
E – Falcon, American peregrine
E – Plover, piping
E – Tern, roseate
E – Turtle, hawksbill sea
E – Turtle, Kemp's
E – Turtle, leatherback sea

Plant

E – Sandplain gerardia

SOUTH CAROLINA

Animals

E – Bat, Indiana
E – Heelsplitter, Carolina
E – Manatee, West Indian (Florida)
E – Plover, piping
E – Stork, wood
E – Tern, roseate
E – Turtle, hawksbill sea
E – Turtle, Kemp's (Atlantic) ridley sea
E – Turtle, leatherback sea
E – Woodpecker, red-cockaded

Plants

E – American chaffseed
E – Black-spored quillwort

E – Bunched arrowhead
E – Canby's dropwort
T – Dwarf-flowered heartleaf
E – Harperella
E – Michaux's sumac
E – Mountain sweet pitcher-plant
E – Persistent trillium
E – Pondberry
E – Relict trillium
E – Rough-leaved loosestrife
E – Schweinitz's sunflower
E – Smooth coneflower

SOUTH DAKOTA

Animals

E – Beetle, American burying
    (giant carrion)
E – Crane, whooping
E – Curlew, Eskimo
E – Ferret, black-footed
E – Plover, piping
E – Sturgeon, pallid
E – Tern, least
E – Wolf, gray

Plants

(No plants on the endangered list)

TENNESSEE

(Tennessee has 81 plant and animal species that are listed as endangered or threatened. The following list is only a selection of those plants and animals that are endangered. Contact the U.S. Fish and Wildlife Service to see the entire list.)

Animals

E – Bat, gray
E – Bat, Indiana
E – Combshell, upland
E – Crayfish, Nashville
E – Darter, amber
E – Fanshell
E – Lampmussel, Alabama
E – Madtom, Smoky
E – Marstonia (snail), (royalobese)
E – Moccasinshell, Coosa
E – Mussel, ring pink (golf stick pearly)
E – Pearlymussel, Appalachian monkeyface

E - Pearlymussel, Cumberland bean
E - Riversnail, Anthony's
E - Spider, spruce-fir moss
E - Squirrel, Carolina northern flying
E - Sturgeon, pallid
E - Tern, least
E - Wolf, red
E - Woodpecker, red-cockaded

Plants

E - Cumberland sandwort
E - Green pitcher-plant
E - Large-flowered skullcap
E - Leafy prairie-clover (Dalea)
E - Roan Mountain bluet
E - Rock cress
E - Rock gnome lichen
E - Ruth's golden aster
E - Spring Creek bladderpod
E - Tennessee purple coneflower
E - Tennessee yellow-eyed grass

TEXAS

(Texas has 70 plant and animal species that are listed as endangered or threatened. The following list is only a selection of those plants and animals that are endangered. Contact the U.S. Fish and Wildlife Service to see the entire list.)

Animals

E - Bat, Mexican long-nosed
E - Beetle, Coffin Cave mold
E - Crane, whooping
E - Curlew, Eskimo
E - Darter, fountain
E - Falcon, northern aplomado
E - Jaguarundi
E - Manatee, West Indian (Florida)
E - Minnow, Rio Grande silvery
E - Ocelot
E - Pelican, brown
E - Plover, piping
E - Prairie-chicken, Attwater's greater
E - Pupfish, Comanche Springs
E - Salamander, Texas blind
E - Spider, Tooth Cave
E - Tern, least
E - Toad, Houston
E - Turtle, hawksbill sea

E - Turtle, Kemp's (Atlantic) ridley sea
E - Vireo, black-capped
E - Warbler, golden-cheeked
E - Woodpecker, red-cockaded

Plants

E - Ashy dogweed
E - Black lace cactus
T - Hinckley's oak
E - Large-fruited sand-verbena
E - Little Aguja pondweed
E - Lloyd's hedgehog cactus
E - Nellie cory cactus
E - Sneed pincushion cactus
E - South Texas ambrosia
E - Star cactus
E - Terlingua Creek cats-eye
E - Texas poppy-mallow
E - Texas snowbells
E - Texas wild-rice
E - Tobusch fishhook cactus
E - Walker's manioc

UTAH
Animals

E - Ambersnail, Kanab
E - Chub, bonytail
E - Chub, humpback
E - Chub, Virgin River
E - Crane, whooping
E - Ferret, black-footed
E - Flycatcher, Southwestern willow
E - Snail, Utah valvata
E - Squawfish, Colorado
E - Sucker, June
E - Sucker, razorback
E - Woundfin

Plants

E - Autumn buttercup
E - Barneby reed-mustard
E - Barneby ridge-cress (peppercress)
E - Clay phacelia
E - Dwarf bear-poppy
E - Kodachrome bladderpod
E - San Rafael cactus
E - Shrubby reed-mustard (toad-flax cress)
E - Wright fishhook cactus

VERMONT

Animals

E – Bat, Indiana
E – Mussel, dwarf wedge

Plants

E – Jesup's milk-vetch
E – Northeastern (Barbed bristle) bulrush

VIRGINIA

Animals

E – Bat, gray
E – Bat, Indiana
E – Bat, Virginia big-eared
E – Darter, duskytail
E – Falcon, American peregrine
E – Fanshell
E – Isopod, Lee County cave
E – Logperch, Roanoke
E – Mussel, dwarf wedge
E – Pearlymussel, Appalachian monkeyface
E – Pearlymussel, birdwing
E – Pearlymussel, cracking
E – Pearlymussel, Cumberland monkeyface
E – Pearlymussel, dromedary
E – Pearlymussel, green-blossom
E – Pearlymussel, little-wing
E – Pearlymussel, pink mucket
E – Pigtoe, fine-rayed
E – Pigtoe, rough
E – Pigtoe, shiny
E – Plover, piping
E – Riffleshell, tan
E – Salamander, Shenandoah
E – Snail, Virginia fringed mountain
E – Spinymussel, James River (Virginia)
E – Squirrel, Delmarva Peninsula fox
E – Squirrel, Virginia northern flying
E – Turtle, hawksbill sea
E – Turtle, Kemp's (Atlantic) ridley sea
E – Turtle, leatherback sea
E – Woodpecker, red-cockaded

Plants

E – Northeastern (Barbed bristle) bulrush
E – Peter's Mountain mallow
E – Shale barren rock-cress
E – Smooth coneflower

WASHINGTON

Animals

E – Caribou, woodland
E – Deer, Columbian white-tailed
E – Pelican, brown
E – Turtle, leatherback sea
E – Wolf, gray

Plants

E – Bradshaw's desert-parsley
     (lomatium)
E – Marsh sandwort

WEST VIRGINIA

Animals

E – Bat, Indiana
E – Bat, Virginia big-eared
E – Clubshell
E – Falcon, American peregrine
E – Fanshell
E – Mussel, ring pink
E – Pearlymussel, pink mucket
E – Pearlymussel, tubercled-blossom
E – Riffleshell, northern
E – Spinymussel, James River
E – Squirrel, Virginia northern flying

Plants

E – Harperella
E – Northeastern (Barbed bristle)
     bulrush
E – Running buffalo clover
E – Shale barren rock-cress

WISCONSIN

Animals

E – Butterfly, Karner blue
E – Dragonfly, Hine's emerald
E – Mussel, winged mapleleaf
E – Pearlymussel, Higgins' eye
E, T – Plover, piping
E – Warbler, Kirtland's
E – Wolf, gray

Plants

(No plants on endangered list)

WYOMING

Animals

E – Crane, whooping
E – Dace, Kendall Warm Springs
E – Ferret, black-footed
E – Squawfish, Colorado
E – Sucker, razorback
E – Toad, Wyoming
E – Wolf, gray

Plants

(No plants on endangered list)

APPENDIX C: WEBSITES BY CLASSIFICATION

Please note that the authors have made a consistent effort to include up-to-date Websites. However, over time, some Websites may move or no longer be posted.

ACID MINE DRAINAGE

National Reclamation Center, West Virginia University, Evansdale office, http//www.nrcce.wvu.edu/

ACID RAIN

http://www.epa.gov/docs/acidrain/andhome/html.

The EPA has a hotline to request educational materials or respond to questions regarding acid rain: (202) 343–9620. http://www.econet.apc.org/acid rain.

Environmental Protection Agency, http://www.epa.gov/docs/acidrain/effects/enveffct.html.

National Reclamation Center's West Virginia University, Evansdale office: http://www.nrcce.wvu.edu/

USGS Water Science/Acid Rain, http://wwwga.usgs.gov/edu/acidrain.html.

AGENCY FOR TOXIC SUBSTANCES AND DISEASES

Registry Division of Toxicology
1600 Clifton Road NE Mailstop E-29
Atlanta, GA 30333
Website: http://www.atsdr1.atsdr.cdc.gov:8080/atsdrhome.html.

Agency for Toxic Substances and Disease Registry, http://www.atsdr.cdc.gov/cxcx3.html.

Information on biosphere reserves and UNESCO's Man and the Biosphere Programme, UNESCO: http://www.unesco. org

Man and the Biosphere Program: http://www. mabnet.org

AGRICULTURE

United States Department of Agriculture, http://www.usda.gov.

ALTERNATIVE FUELS

Department of Energy, http://www.doe.gov.

Department of Energy Alternative Fuels Data Center, http://www.afdc.nrel.gov; http://www.afdc.doe.gov/; or http://www.fleets.doe.gov.

AMPHIBIANS

http://www.frogweb.gov/

ANTARCTICA

Antarctica Treaty, http://www.sedac.ciesin.org/pidb/register/reg-024.rrr.html.

Greenpeace International Antarctic Homepage, http://www.greenpeace.org/~comms/98/antarctic.

International Centre for Antarctic Information and Research Homepage (includes text of Antarctic Treaty), http://www.icair.iac.org.nz.

Virtual Antarctica, http://www.exploratorium.edu

ARCTIC

Arctic Circle (University of Connecticut), http://arcticcircle.uconn.edu/arcticcircle.

Arctic Council Home Page, http://www.nrc.ca/arctic/index.html.

Arctic Monitoring and Assessment Programme (Norway), http://www.gsf.de/ UNEP/amap1.html.

Arctic National Wildlife Refuge, http://energy.usgs.gov/factsheets/ANWR/ANWR.html.

Institute of Arctic and Alpine Research, http://instaar.colorado.edu.

Institute of the North (Alaska Pacific University),

Inuit Circumpolar Conference,
NOAA Fisheries, http://www.nmfs.gov/.

Nunavut,
Smithsonian Institution Arctic Studies Center,
http://www.mnh.si.edu/arctic.

U.S. Fish and Wildlife Service
U.S. Department of the Interior

1849 C Street, NW,
Washington, D.C. 20240
Telephone: (202) 208-5634
Website: http://www.fws.gov.

World Conservation Monitoring Centre Arctic
Programme, http://www.wcmc.org.uk/
arctic.

AUTOMOBILE

Cars and Their Enviromental Impact,
http://www.environment.volvocars.com/
ch1-1.htm.

National Center for Vehicle Emissions Control
and Safety (NCVECS), http://www.colostate.
edu/Depts/NCVECS/ncvecs1.html.

U.S. Environmental Protection Agency Fact Sheet
(EPA 400-F-92-004, August 1994), "Air Toxics
from Motor Vehicles," http://www.epa.gov/
oms/02-toxic.htm.

U.S. Enviromental Protection Agency,
Office of Mobile Sources,
http://www.epa.gov/oms.

BIOLOGICAL WEAPONS

Federation of American Scientist Biological
Weapons Control, http://www.fas.org/bwc.

Chemical and Biological Defense Information
Analysis Center, http://www.cbiac.apgea.
army. mil

BIOMES

Committee for the National Institute for the
Environment, http://www.cnie.org/nle/
biodv-6.html.

BIOREMEDIATION

Consortium, http://www.rtdf.org/public/
biorem.

BROWNFIELD

Projects, http://www.epa.gov/brownfields/.

CERES

Website: http://www.ceres.org or
e-mail ceres@igc.apc.org.
Summaries of Major Environmental Laws,
http://www.epa.gov/region5/defs/index.html.

CHEETAHS

Cheetah Conservation Fund

4649 Sunnyside Avenue N, Suite 325
Seattle, WA 98103
Website: http://www.cheetah.org.

World Wildlife Fund

1250 24th Street, NW,
Washington, D.C. 20037
Telephone: 1-800-225-5993
Website: http://www.worldwildlife.org/.

CHEMICAL WEAPONS

Chemical Stockpile Disposal Project (CSDP),
http://www.pmcd.apgea.army.mil/
graphical/CSDP/index.html.

Tooele Chemical Agent Disposal Site Facility,
http://www.deq.state.ut.us/eqshw/cds/
tocdfhp1.htm.

CLEAN WATER ACT

Sierra Club, "Happy 25th Birthday, Clean
Water Act," http://sierraclub.org/wetlands/
cwabday.html.

CLIMATE CHANGE AND
GLOBAL WARMING

U.S. Geological Survey, Climate Change and
History, http://geology.usgs.gov/index.shtml.

EPA Global Warming Site,
http://www.epa.gov/globalwarming.

Greenpeace International, Climate,
http://www.greenpeace.org/~climate.

United Nations Intergovernmental Panel on
Climate Change, http://www.ipcc.ch.

COAL

Coal Age Magazine, http://coalage.com.

Department of Energy, Office of Fossil Energy, http:/www.doe.gov.

U.S. Geological Survey, National Coal Resources Data System, http:energy.er.usgs.gov/ coalqual. htm.

COASTAL AND MARINE GEOLOGY

U.S. Geological Survey, http://marine.usgs.gov/.

COMPOSTING

EPA Office of Solid Waste and Emergency Response—Composting, http:www.epa. gov/epaoswer/non-hw/compost/index.htm
Cornell Composting, http://www.cfe.cornell. edu/compost/Composting_Homepage.html

CONSENT DECREES

EPA Office of Enforcement and Compliance Assurance, http://es.epa.gov/oeca/osre/ decree.html.

CORAL REEFS

Coral Reef Alliance, http://www.coral.org.

Coral Reef Network Directory, Greenpeace
1436 U Street, NW
Washington, D.C. 20009
Website: http://www.greenpeace.org.

EARTHDAY 2000

Earth Day Network

91 Marion Street,
Seattle, WA 98104
Telephone: 1(206)-264-0114.
Website: http://www.earthday.net/;
and worldwide@earthday.net.

EARTHWATCH

Earthwatch Institute International,
http://www. earthwatch.org.

EL NIÑO

El Niño/La Niña theme page, contact NOAA
Website: http://www.pmel.noaa.gov/toga-tao/
el-nino/nino-home-low.html.

NOAA, La Niña homepage, www.elnino.noaa.
gov/lanina.html.

National Center for Atmospheric Research,
http://www.ncar.ucar.edu/.

National Hurricane Center/Tropical Prediction Center, http://www.nhc.noaa.gov/.

National Oceanographic and Atmospheric
Administration, http://www.noaa.gov/.

Scripps Institute of Oceanography,
http://sio.ucsd.edu/supp_groups/siocomm/
elnino/elnino.html.

ELECTRIC VEHICLES

Electric Vehicle Association of the Americas
800-438-3228, http://www.evaa.org.

Electric Vehicle Technology, http://www.avere.org/.

ELEPHANTS

African Wildlife Foundation,
http://www.awf.org.

U.S. Fish and Wildlife Service, Species List of Endangered and Threatened Wildlife,
http://endangered.fws.gov/

World Wildlife Fund, http://www.wwf.org.

ETHANOL

U.S. Department of Energy, Energy Efficiency and Renewable Energy Clearinghouse,

P.O. Box 3048
Merrifield, VA 22116
E-mail: energyinfo@delphi.com.
Website: http://www.doe.gov.

EVERGLADES

National Park Service, Everglades National Park,
http://www.nps.gov/ever.

FEDERAL EMERGENCY MANAGEMENT AGENCY (FEMA)

FEMA, http://www.fema.gov.

FISHING, COMMERCIAL

National Oceanographic and Atmospheric
Administration Fisheries,
http://www.nmfs. gov/.

United Nations Food and Agriculture Organization Fisheries, http://www.fao.org/waicent/faoinfo/fishery/fishery.htm.

FORESTS

American Forests, http://www.amfor.org.

Greenpeace International, Forests, http://www. greenpeace.org/~forests.

Society of American Foresters, http://www. safnet.org.

U.S. Forest Service, http://www.fs.fed.us.

U.S. Forest Service Research, http://www.fs.fed.us/links/research.shtml.

World Conservation Monitoring Centre, http://www.wcmc.org.uk.

World Resources Institute Forest Frontiers Initiative, http://www.wri.org/ffi.

World Wildlife Fund (Worldwide Fund for Nature) Forests for Life Campaign, http://www.panda.org/forests4life.

FUEL CELLS AND OTHER ALTERNATIVE FUELS

Crest's Guide to the Internet's Alternative Energy Resources, http://solstice.crest.org/online/aeguide/aehome.html.

U.S. Department of Energy

P.O. Box 12316
Arlington, VA 22209
Telephone: 1-800-423-1363
Website: http://www.doe.gov.

U.S. Department of Energy, Alternative Fuels Data Center, http://www.afdc.nrel.gov.

GEOLOGY

Geological surveys, U.S. Geological Survey, http://www.usgs.gov/.

For general interest publications and products, http://mapping.usgs.gov/www/products/mappubs.html.

GEOTHERMAL SITES

Energy and Geoscience Institute

University of Utah
423 Wakara Way

Salt Lake City, UT 84108
Website: http://www.egi.utah.edu.

Geothermal energy information, http://geothermal.marin.org.

Geothermal database USA and Worldwide, http://www.geothermal.org.

International geothermal, http://www.demon.co.uk/geosci/igahome.html.

Solstice is the Internet information service of the Center for Renewable Energy and Sustainable Technology (CREST), http://solstice. crest.org/

GLACIERS SHRINKING

United States Geological Survey, Climate Change and History, http://geology.usgs.gov/index. shtml.

Sierra Club, Public Information Center, (415) 923-5653; or the Global Warming and Energy Team, (202) 547-1141, or by E-mail: information@sierraclub.org.

GLOBEC

Educational Website, http://cbl.umces.edu/fogarty/usglobec/misc/education.html.

GRASSLANDS AND PRAIRIES

Postcards from the Prairie, http://www.nrwrc.usgs.gov/postcards/postcards.htm.

University of California, Berkeley, World Biomes, Grasslands, http://www.ucmp.berkeley.edu/glossary/gloss5/biome/grasslan.html.

Worldwide Fund for Nature, Grasslands and Its Animals, http://www.panda.org/kids/wildlife/idxgrsmn.htm.

GROUNDWATER

EPA, http://www.epa.gov/swerosps/ej/.

Groundwater atlas of the United States, http://www.capp.er.usgs.gov/publicdocs/gwa/.

HAZARDOUS MATERIALS TRANSPORTAION ACT

Website: http://www.dot.gov.

HAZARDOUS SUBSTANCES

U.S. Environmental Protection Agency Program, http://epa.gov/.

U.S. Occupational Safety and Health Administration (OSHA), http://www.osha.gov/toxicsubstances/index.html.

Environmental Defense Fund (data on wastes and chemicals at U.S. sources), http://www.scorecard.org.

HAZARDOUS WASTE TREATMENT

Federal Remedial Technologies Roundtable, Hazardous Waste Clean-Up Information ("CLU-IN"), http://www.clu-in.org.

HEAVY METALS

U.S. Environmental Protection Agency, Office of Pollution Prevention and Toxics, http://www.epa.gov/opptintr.

HIGH-LEVEL RADIOACTIVE WASTES

U.S. Nuclear Regulatory Commission, Radioactive Waste Page, http://www.nrc.gov/NRC/radwaste.

U.S. Environmental Protection Agency, Mixed-Waste Homepage, http://www.epa.gov/radiation/mixed-waste.

HURRICANES

National Hurricane Center, http://www.nhc.noaa.gov.

HYDROELECTRIC POWER

U.S. Bureau of Reclamation Hydropower Information, http://www.usbr.gov/power/edu/edu.htm.

U.S. Geological Survey, http://wwwga.usgs.gov/edu/hybiggest.html.

HYDROGEN

National Renewable Energy Laboratory, http://www.nrel.gov/lab/pao/hydrogen.html.

EnviroSource, Hydrogen InfoNet, http:///www.eren.doe.gov/hydrogen/infonet.html.

INTERNATIONAL ATOMIC ENERGY AGENCY

Agency, http://www.iaea.org.

Managing Radioactive Waste Fact Sheet, http://www.iaea.org/worldatom/inforesource/factsheets/manradwa.html.

INTERNATIONAL COUNCIL FOR LOCAL ENVIRONMENTAL INITIATIVES

Homepage, http://www.iclei.org.

INTERNATIONAL REGISTER OF POTENTIALLY TOXIC CHEMICALS

Homepage, http://www.unep.org/unep/program/hhwb/chemical/irptc/home.htm.

INTERNATIONAL WHALING COMMISSION

Homepage, http://www.ourworld.compuserve.com/homepages/iwcoffice.

INVERTEBRATES: THREATENED AND ENDANGERED

U.S. Fish and Wildlife Service, Species List of Endangered and Threatened Wildlife, http://endangered.fws.gov/

LANDSAT AND SATELLITE IMAGES

Earthshots, Satellite Images of Environmental Change, http://www.usgs.gov/Earthshots/.

Landsat Gateway, http://landsat.gsfc.nasa.gov/main.htm.

LEAD

National Lead Information Center's Clearinghouse, 1-800-424-LEAD, http://www.epa.gov/lead/.

LEOPARDS

U.S. Fish and Wildlife Service, Species List of Endangered and Threatened Wildlife, http://www.fws.gov/r9endspp/lsppinfo.html.

LITTER

Keep America Beautiful, http://www.kab.org.

MAMMALS

U.S. Fish and Wildlife Service, Vertebrate Animals, http://www.fws.gov/r9endspp/lsppinfo.html.

MANATEES

Save the Manatees, http://www.savethemanatee. org.

Sea World, Manatees, http://www.seaworld.org/manatee/sciclassman.html.

MARSHES

Environmental Protection Agency, Office of Wetlands, Oceans, Watersheds, http://www.epa.gov/owow/wetlands/wetland2.html.

North American Waterfowl and Wetlands Office, http://www.fws.gov/r9nawwo.

North American Wetlands Conservation Act, http://www.fws.gov/r9nawwo/nawcahp.html.

North American Wetlands Conservation Council, http://www.fws.gov/r9nawwo/nawcc.html.

Wetlands, wetlands-hotline@epamail.epa.gov.

MATERIAL SAFETY DATA SHEET

Toxic chemicals, http://www.siri.org/msds; http://www.ilpi.com/mads/index.html.

MENDES, CHICO

Chico Mendes, http://www.edf.org/chico.

NATURAL DISASTERS

Building Safer Structures, http://quake.wr.usgs. gov/QUAKES/FactSheets/SaferStructures/.

Center for Integration of Natural Disaster Information, http://cindi.usgs.gov/events/.

Earthquakes, http://quake.wr.usgs.gov/; http://geology.usgs.gov/quake.html. For the latest earthquake information http://quake.wr.usgs.gov/QUAKES/CURRENT/current.html

National Hurricane Center, http://www.nhc.noaa.gov.

U.S. Geological Survey, http://geology.usgs.gov/whatsnew.html.

NATIONAL MARINE FISHERIES

History of National Marine Fisheries Service, http://www.wh.whoi.edu/125th/history/century.html.

National Marine Fisheries, http://kingfish.ssp.nmfs.gov.

NOAA Fisheries, http://www.nmfs.gov/.

NATIONAL OCEAN AND ATMOSPHERIC ADMINISTRATION (NOAA)

Climate forecasting, http://www.cdc.noaa.gov/ Seasonal/.

El Niño Theme Page, http://www.pmel.noaa.gov/toga-tao/el-nino/nino-home-low.html.

Homepage, http://www.noaa.gov/.

Recover Protected Species, http://www.noaa.gov/nmfs/recover.html.

Safe Navigation Page, http://anchor.ncd.noaa.gov/psn/psn.htm.

NATIONAL WEATHER SERVICE

Homepage, http://www.nws.noaa.gov.

NATIONAL WILDLIFE REFUGE SYSTEM

Homepage, http://refuges.fws.gov/NWRSHomePage.html.

NATURAL GAS

American Gas Association, http://www.aga.org.

Oil and Gas Journal Online, http://www.ogjonline.com.

U.S. Department of Energy, Energy Information Administration, http://www.eia.doe.gov.

U.S. Department of Energy, Office of Fossil Energy, http://www.fe.doe.gov.

U.S. Geological Survey Energy, Resources Program, http://energy.usgs.gov/index.html.

NOISE POLLUTION

Noise Pollution Clearinghouse, http://www. nonoise.org.

NONPOINT SOURCES

Nonpoint Source Pollution Control Program, http://www.epa.gov/OWOW/NPS/ whatudo.html; http://www.epa.gov/ OWOW/ NPS/.

NUCLEAR ENERGY AND NUCLEAR REACTORS

American Nuclear Society, http://www.ans.org.

Nuclear Energy Institute, http://www.nei.org.

Nuclear Information and Resource Service, http://www.nirs.org.

U.S. Department of Energy, Office of Nuclear Energy, Science and Technology, http://www.ne.doe.gov.

U.S. Nuclear Regulatory Commission, http://www.nrc.gov.

NUCLEAR WASTE POLICY ACT

American Nuclear Society, http://www.ans.org.

Nuclear Energy Institute, http://www.nei.org.

NUCLEAR WASTE SITES

Hazard Ranking System, http://www.epa. gov/ superfund/programs/npl_hrs/ hrsint.htm.

National Research Council, Board on Radioactive Waste Management, http://www4.nas.edu/ brwm/brwm-res.nsf.

Superfund, http://www.pin.org/superguide.htm; http://www.epa.gov/superfund.

U.S. Department of Energy, Office of Civilian Radioactive Waste Management, http://www.rw.doe.gov.

U.S. Environmental Protection Agency, Mixed-Waste Homepage, http://www.epa. gov/radiation/mixed-waste.

U.S. Nuclear Regulatory Commission, Radioactive Waste Page, http://www.nrc.gov/ NRC/ radwaste.

OCCUPATIONAL SAFETY AND HEALTH ACT (OSHA)

OSHA Homepage, http://www.osha.gov.

OCEAN THERMAL ENERGY CONVERSION (OTEC)

National Renewable Energy Laboratory

1617 Cole Boulevard
Golden, CO 80401
Website: http:llnrelinfo.nrel.gov.

Natural Energy Laboratory of Hawaii, http://bigisland.com/nelha/index.html.

OCEANS

National Oceanographic and Atmospheric Administration, http://www.noaa.gov/.

Safe Ocean Navigation Page, http://anchor.ncd. noaa.gov/psn/psn.htm.

OFFICE OF SURFACE MINING

Office of Surface Mining, http://www.osmre.gov.

Appalachian Clean Streams Initiative, majordomo@osmre.gov.

OLD-GROWTH FORESTS

Greenpeace International, Forests, http://www.greenpeace.org/~forests.

World Resources Institute, Forest Frontiers Initiative, http://www.wri.org/ffi.

OLMSTEAD, FREDERICK LAW

Homepage, http://fredericklawolmsted.com.

ORGANIZATION OF PETROLEUM EXPORTING COUNTRIES (OPRC)

Homepage, http://www.opec.org.

OVERFISHING

Information and data statistics, http://www.nmfs. gov.

National Aeronautics and Space Administration, Ocean Planet, http://seawifs.gsfc.nasa.gov/ OCEAN_PLANET/HTML/ peril_overfishing.html.

National Marine Fisheries Service, http://www. nmfs.gov.

NOAA, http://www.noaa.gov.

United Nations Food and Agricultural Organization, http://www.fao.org.

United Nations Food and Agriculture Organization Fisheries, http://www.fao.org/.

United Nations System, http://www.unsystem.org.

OZONE-RELATED ISSUES

Environmental Protection Agency, science of ozone depletion, http://www.epa.gov/ozone/science/.

NOAA, Commonly Asked Questions about Ozone, www.publicaffairs.noaa.gov/grounders/ozo1.html.

NOAA, Network for the Detection of Stratospheric Change, www.noaa.gov.

PARROTS

Online Book of Parrots, http://www.ub.tu-clausthal.dep/p_welcome.html.

World Parrot Trust, http://www.worldparrottrust.org.

World Wildlife Fund, http:www.panda.org.

PESTICIDES

Toxics and Pesticides, http://www.epa.gov/oppfead1/work_saf/.

Pesticides in the Atmosphere, http://ca.water.usgs.gov/pnsp/atmos.

PETERSON, ROGER TORY

Roger Tory Peterson Institute of Natural History,

311 Curtis Street
Jamestown, NY 14701
Website: http://www.rtpi.org/info/rtp.htm.

PETROLEUM

American Petroleum Institute, http://www.api.org.

Petroleum Information, http://www.petroleuminformation.com.

Oil and Gas Journal Online, http://www. ogjonline.com.

U.S. Department of Energy, Energy Information Administration, http://www.eia.doe.gov.

U.S. Department of Energy, Office of Fossil Energy, http://www.fe.doe.gov.

U.S. Geological Survey Energy Resources Program, http://energy.usgs.gov/index.html.

U.S. Geological Survey Fact Sheet FS-145-97, "Changing Perceptions of World Oil and Gas Resources as Shown by Recent USGS Petroleum Assessments," http://greenwood.cr.usgs.gov/pub/fact-sheets/fs-0145-97/fs-0145-97.html.

PLUTONIUM

U.S. Nuclear Regulatory Commission, Radioactive Waste Page, http://www.nrc.gov/NRC/radwaste.

RADIATION AND RADIOACTIVE WASTES

International Atomic Energy Agency, "Managing Radioactive Waste" Fact Sheet, http://www.iaea.org/worldatom/inforesource/factsheets/manradwa.html.

National Research Council, Board on Radioactive Waste Management, http://www4.nas.edu/brwm/brwm-res.nsf.

U.S. Department of Energy, Office of Civilian Radioactive Waste Management, http://www. rw.doe.gov.

U.S. Environmental Protection Agency, Mixed-Waste Homepage, http://www.epa.gov/radiation/mixed-waste.

U.S. Nuclear Regulatory Commission, Radioactive Waste Page, http://www.nrc.gov/NRC/radwaste.

RADON

Radon in Earth, Air, and Water, http://sedwww.cr.usgs.gov:8080/radon/radonhome.html.

RAIN FORESTS

Greenpeace International, forests, http://www.greenpeace.org/~forests.

Rainforest Action Network (RAN)

President Randy Hayes
221 Pine Street Suite 500
San Francisco, CA 94104
Telephone: (415) 398-4404
Website: http://www.ran.org

Rainforest Alliance (RA)

65 Bleeker Street
New York, NY 10012
Website: http://www.rainforest-alliance.org

U.S. Forest Service, http://www.fs.fed.us.

World Wildlife Fund (Worldwide Fund for
Nature), Forests for Life Campaign,
http://www.panda.org/forests4life.

RESOURCE CONSERVATION AND RECOVERY ACT

Homepage, http://www.epa.gov/epaoswer/hotline.

SALMON

National Marine Fisheries Service, http://www.
nwr.noaa.gov/1salmon/salmesa/index.htm.
NOAA Fisheries, http://www.nmfs.gov/.

SALT MARSHES

National Wetlands Research Center,
http://www.nwrc.usgs.gov/educ_out.html.

USGS Coastal and Marine Geology,
http://marine.usgs.gov/.

SANITARY LANDFILLS

Solid waste management, http://web.mit.edu/
urbanupgrading/urban environment/

*Landfills - Solid and Hazardous Waste and Ground-
water Quality Protection*, http://www.gfredlee.
com/plandfil2.htm

SIBERIA

Siberia, http://www.cnit.nsk.su/univer/english/
siberia.htm.

SOLAR ENERGY

American Solar Energy Society

2400 Central Avenue, Suite G-1
Boulder, CO 80301.
Website: http://www.soton.ac.uk/~solar/.

Solar Energy Industries Association

122 C Street, NW, 4th Floor
Washington, D.C. 20001.
Website: http://www.seia.org/main.htm.

U.S. Department of Energy, Photovoltaic Program,
http://www.eren.doe.gov/pv/text_frameset.
html.

SOLAR POND

Department of Mechanical and Industrial Engineering

University of Texas at El Paso
El Paso, TX 79968.
E-mail: aswift@cs.utep.edu.

SPENT FUEL

Environmental Protection Agency, www.ntp.doe.
gov, www.rw.doe.gov/pages/resource/facts/
transfct.htm.

SUPERFUND

Environmental Protection Agency,
http://www.epa.gov/epaoswer/hotline.

Recycled Superfund sites, http://www.epa.gov/
superfund/programs/recycle/index.htm.

Superfund Information, http://www.epa.gov/
superfund.

U.S. EPA Superfund Program Homepage,
Website: http://www.epa.gov/superfund/
index.htm.

TENNESSEE VALLEY AUTHORITY

Homepage, http://www.tva.gov.

THOREAU, HENRY

Website: http://www.walden.org.

TOXIC CHEMICALS

Environmental Defense Fund,
http://www.scorecard.org.

U.S. Department of Health and Human
Services, Agency for Toxic Substances and
Disease Registry (ASTDR),
http://www.atsdr.cdc.gov/

U.S. Environmental Protection Agency, Integrated
Risk Information System (IRIS),
http://www. siri.org/msds;
http://www.ilpi.com/mads/index.html.

U.S. Occupation Health and Safety Administration, http://www.toxicsubstances/index.html.

TOXIC RELEASE INVENTORY

Environmental Defense Fund, http://www.scorecard.org.

Environmental Protection Agency, http://www.epa.gov.

Teach with Databases, Toxic Release Inventory, http://www.nsta.org/pubs/special/pb143x01.htm.

TOXIC WASTE

Environmental Defense Fund, http://www.scorecard.org.

Institute for Global Communications, http://www.igc.org/igc/issues/tw/.

TRADE RECORDS ANALYSIS OF FLORA AND FAUNA IN COMMERCE (TRAFFIC)

Homepage, http://www.traffic.org/about/.

URBAN FORESTS

American Forests, http://www.amfor.org.

TreeLink, http://www.treelink.org.

VERTEBRATES

U.S. Fish and Wildlife Service, Species List of Endangered and Threatened Wildlife, http://www.fws.gov/r9endspp/lsppinfo.html.

VICUNA

U.S. Fish and Wildlife Service, Species List of Endangered and Threatened Wildlife, http://endangered.fws.gov

VITRIFICATION

U.S. Department of Energy, http://www.em.doe.gov/fs/fs3m.html.

VOLCANOES

USGS, Volcanoes in the Learning Web, http://www.usgs.gov/education/learnweb/volcano/index.html.

Volcano Hazards, http://volcanoes.usgs.gov/.

WATER CONSERVATION AND POLLUTION

Early History of the Clean Water Act, http://epa.gov/history/topics.

Environmental Protection Agency, Office of Wetlands, Oceans, Watersheds for Nonpoint Source information, http://www.epa.gov/owow/wetlands/wetland2.html; http://www.epa.gov/swerosps/ej/.

U.S. Geological Survey, Water Resources of the United States, National Groundwater Association Homepage, http://www.h2o-ngwa.org.

Water Resources Information, http://water.usgs.gov/.

Water Use Data, http://water.usgs.gov/public/watuse/.

WETLANDS

National Wetlands Research Center, http://www.nwrc.usgs.gov/educ_out.html.

Ramsar Convention on Wetlands (International), http://www2.iucn.org/themes/ramsar/.

Ramsar List of Wetlands of International Importance, http://ramsar.org/key_sitelist.htm.

WHALES

Institute of Cetacean Research (ICR), http://www.whalesci.org.

U.S. Fish and Wildlife Service, Species List of Endangered and Threatened Wildlife, http://www.fws.gov/r9endspp/lsppinfo.html; http://www.highnorth.no/iceland/th-in-to.htm; http://greenpeace.org/.

WILDERNESS

U.S. Forest Service, *Roadless Area Review and Evaluation*, http://www.fs.fed.us.

Wilderness Society, http://www.wilderness.org/newsroom/factsheets.htm.

WILDLIFE REFUGES

Conservation International, http://www.conservation.org.

Nature Conservancy, http://www.tnc.org.

U.S. Fish and Wildlife Service, National Wildlife Refuge System, http://refuges.fws.gov.

World Conservation Union/International Union for the Conservation of Nature, http://www.iucn.org.

WIND ENERGY

American Wind Energy Association

122 C Street NW, 4th Floor
Washington, D.C. 20001
Telephone: (202) 383-2500.
E-mail: awea@mcimail.com.
Website: http://www.awea.org.

Center for Renewable Energy and Sustainable Technology (CREST)

Solar Energy Research and Education Foundation
777 North Capitol Street NE, Suite 805
Washington, D.C. 20002
Website: http://solstice.crest.org/.

WOLVES

U.S. Fish and Wildlife Service, http://www.fws.gov/.
U.S. Fish and Wildlife Service, Species List of Endangered and Threatened Wildlife, http://endangered.fws.gov/.

World Wildlife Fund

1250 24th Street, NW
Washington, D.C. 20037
Telephone: 1-800-225-5993
Website: http://www.worldwildlife.org/.

WORLD HEALTH ORGANIZATION

Homepage, http://www.who.int.

WORLD WILDLIFE FUND

1250 24th Street, NW
Washington, D.C. 20037
Telephone: 1-800-225-5993
Website: http://www.wwf.org/.

YUCCA MOUNTAIN PROJECT

Homepage, http://www.ymp.gov/.

ZEBRAS

U.S. Fish and Wildlife Service, Species List of Endangered and Threatened Wildlife, http://endangered.fws.gov/.

ZOOS

Bronx Zoo, http://www.bronxzoo.com/.
San Diego Zoo, http://www.sandiegozoo.org/.

Appendix D: Environmental Organizations

Action for Animals

P.O. Box 17702
Austin, TX 78760
Telephone: (512) 416-1617
Fax: (512) 445-3454
Website: http://www.envirolink.org/

African Wildlife Foundation (AWF)

1400 Sixteenth Street, NW, Suite 120
Washington, D.C. 20036
Telephone: (202) 939-3333
Fax: (202) 939-3332
Website: http://www.awf.org/home.html

Agency for Toxic Substances and Diseases, Registry Division of Toxicology (ATSDR)

1600 Clifton Road
NE Mailstop E-29
Atlanta, GA 30333
Telephone: (888) 42-ATSDR or (888) 422-8737
E-mail: ATSDRIC@cdc.gov
Website: http://www.atsdr.cdc.gov/
contacts.html

Alaska Forum for Environmental Responsibility

P.O. Box 188
Valdez, AK 99686
Telephone: (907) 835-5460
Fax: (907) 835-5410
Website: http://www.accessone.com/~afersea

American Conifer Society (ACS)

P.O. Box 360
Keswick, VA 22947-0360
Telephone: (804) 984-3660
Fax: (804) 984-3660

E-mail: ACSconifer@aol.com
Website: http://www.pacificrim.net/~bydesign/
acs.html

American Forests

P.O. Box 2000
Washington, D.C. 20013
Telephone: (202) 955-4500
Website: http://www.americanforests.org

American Nuclear Society

555 North Kensington Avenue
La Grange Park, IL 60525
Telephone: (708) 352-6611
Fax: (708) 352-0499
E-mail: NUCLEUS@ans.org
Website: http://www.ans.org

American Oceans Campaign

201 Massachusetts Avenue NE, Suite C-3
Washington, D.C. 20002
Telephone: (202) 544-3526
Fax: (202) 544-5625
E-mail: aocdc@wizard.net
Website: http://www.americanoceans.org

American Rivers

1025 Vermont Avenue NW, Suite 720
Washington, D.C. 20005
Telephone: (202) 347-7500
Fax: (202) 347-9240
E-mail: amrivers@amrivers.org
Website: http://www.amrivers.org

American Society for Horticultural Science (ASHS)

600 Cameron Street
Alexandria, VA 22314-2562

Telephone: (703) 836-4606
Fax: (703) 836-2024
E-mail: webmaster@ashs.org
Website: http://www.ashs.org

American Society for the Prevention of Cruelty to Animals (ASPCA)

424 East Ninety-second Street
New York, NY 10128
Telephone: (212) 876-7700
Website: http://www.aspca.org

American Solar Energy Society

2400 Central Avenue, Suite G-1
Boulder, CO 80301
Telephone: (303) 443-3130
Fax: (303) 443-3212
E-mail: ases@ases.org
Website: http://www.ases.org
Publication: *Solar Today*

American Wind Energy Association

122 C Street NW, Fourth Floor
Washington, D.C. 20001
Telephone: (202) 383-2500
E-mail: awea@mcimail.com
Website: http://www.awea.org

Animal Legal Defense Fund (ALDF)

127 Fourth Street
Petaluma, CA 94952
Telephone: (707) 769-7771
Fax: (707) 769-0785
E-mail: info@aldf.org
Website: http://www.aldf.org

Animal Rights Network

P.O. Box 25881
Baltimore, MD 21224
Telephone: (410) 675-4566
Fax: (410) 675-0066
Website: http://www.envirolink.org/arrs/aa/
    index.html
Publication: *Animals' AGENDA*, a bimonthly
    magazine

Baron's Haven Freehold

104 South Main Street
Cadiz, OH 43907

Telephone: (740) 942-8405
Website: http://bhfi.1st.net

Biodiversity Support Program (BSP)

1250 North Twenty-fourth Street NW,
    Suite 600
Washington, D.C. 20037
Telephone: (202) 778-9681
Fax: (202) 861-8324
Website: http://www.BSPonline.org

Biosfera

Pres. Vargas 435, Suites 1104 and 1105
Rio de Janeiro, RJ 20077-900
Brazil

Birds of Prey Foundation

2290 South 104th Street
Broomfield, CO 80020
Telephone: (303) 460-0674
Fax: (303) 666-1050

Build the Earth

3818 Surfwood Road
Malibu, CA 90265
Telephone: (310) 454-0963

Center for Conversion and Research of Endangered Wildlife (CREW)

Cincinnati Zoo and Botanical Garden
3400 Vine Street
Cincinnati, OH 45220
E-mail: terri.roth@cincyzoo.org

Center for Marine Conservation

1725 DeSales Street SW, Suite 600
Washington, D.C. 20036
Telephone: (202) 429-5609
Fax: (202) 872-0619
E-mail: cmc@dccmc.org
Website: http://www.cmc-ocean.org

Centers for Disease Control (CDC)

1600 Clifton Rd.
Atlanta, GA 30333

Telephone: (800) 311-3435
Website: http://www.cdc.gov

Cheetah Conservation Fund (CCF)

P.O. Box 1380
Ojai, CA 93024
Telephone: (805) 640-0390
Fax: (815) 640-0230
E-mail: info@cheetah.org
Website: http://www.cheetah.org

Clean Air Council (CAC)

135 South Nineteenth Street, Suite 300
Philadelphia, PA 19103
Telephone: (888) 567-7796
Website: http://www.libertynet.org/
~cleanair/

Coalition for Economically Responsible Economies (CERES)

11 Arlington Street, Sixth Floor
Boston, MA 02116-3411
Telephone: (617) 247-0700
Fax: (617) 267-5400
Website: http://www.ceres.org

Conservation International

1015 Eighteenth Street NW Suite 1000
Washington, D.C. 20036
Telephone: (202) 429-5660
Website: http://www.conservation.org/
Publication: *Orion Nature Quarterly*

Convention on International Trade in Endangered Species of Wild Fauna and Flora (CITES)

CITES Secretariat
International Environment House,
15, chemin des Anémones, CH–1219
Châtelaine-Geneva, Switzerland
E-mail: cites@unep.ch
Website: http://www.cites.org/index.shtml

Council for Responsible Genetics

5 Upland Road, Suite 3
Cambridge, MA 02140
Website: http://www.gene-watch.org

Cousteau Society

870 Greenbriar Circle, Suite 402
Chesapeake, VA 23320
Telephone: (804) 523-9335
E-mail: cousteau@infi.net
Website: http://www.cousteausociety.org/
Publication: *Calypso Log*

Defenders of Wildlife

1101 Fourteenth Street NW, Room 1400
Washington, D.C. 20005
Telephone: (800) 441-4395
Website: http://www.Defenders.org
Publication: *Defenders*, a quarterly magazine

Dian Fossey Gorilla Fund International

800 Cherokee Avenue SE
Atlanta, GA 30315-1440
Telephone: (800) 851-0203
Fax: (404) 624-5999
E-mail: 2help@gorillafund.org
Website: http://www.gorillafund.org/
000_core_frmset.html

Earth Day Network

1616 P Street NW
Suite 200
Washington, D.C. 20036
E-mail: carthday@carthday.net
Website: http://www.earthday.net

Earth Island Institute (EII)

300 Broadway, Suite 28
San Francisco, CA 94133
Telephone: (415) 788-3666
Fax: (415) 788-7324
Website: http://www.earthisland.org/abouteii/
abouteii.html
Publication: *Earth Island Journal*, a quarterly
magazine

Earth, Pulp, and Paper

P.O. Box 64
Leggett, CA 95585
Telephone: (707) 925-6494
E-mail: tree@tree.org
Website: http://www.tree.org/epp.htm

EarthFirst! (EF!)

P.O. Box 5176
Missoula, MT 59806
Website: http://www.webdirectory.com/
General_Environmental_Interest/
Earth_First_/

Earthwatch Institute

In United States and Canada
3 Clocktower Place, Suite 100
Box 75
Maynard, MA 01754
Telephone: (800) 776-0188 or (617) 926-8200
Fax: (617) 926-8532
In Europe
57 Woodstock Road
Oxford OX2 6HJ, United Kingdom
E-mail: info@uk.earthwatch.org
Website: http://www.earthwatch.org

EcoCorps

1585 A Folsom Avenue
San Francisco, CA 94103
Telephone: (415) 522-1680
Fax: (415) 626-1510
E-mail: eathvoice@ecocorps.org
Website: http://www.owplaza.com/eco

Ecotourism Society

P.O. Box 755
North Bennington, VT 05257
Telephone: (802) 447-2121
Fax: (802) 447-2122
E-mail: ecomail@ecotourism.org
Website: http://www.ecotoursim.org

E. F. Schumacher Society

140 Jug End Road
Great Barrington, MA 01230
Telephone: (413) 528-1737
E-mail: efssociety@aol.com
Website: http://members.aol.com/efssociety/
index.html

Electric Vehicle Association of the Americas

701 Pennsylvania Avenue NW, Fourth Floor
Washington, D.C. 20004

Telephone: (202) 508-5995
Fax: (202) 508-5924
Website: http://www.evaa.org

Environmental Defense Fund (EDF)

257 Park Avenue South
New York, NY 10010
Telephone: (800) 684-3322
Fax: (212) 505-2375
E-mail (for general questions and information):
Contact@environmentaldefense.org
Website: http://www.edf.org
Publication: *Nature Journal*, a monthly
magazine

Exotic Cat Refuge and Wildlife Orphanage

Route 3, Box 96A
Kirbyville, TX 75956
Telephone: (409) 423-4847

Federal Emergency and Management Agency (FEMA)

500 C Street SW
Washington, D.C. 20472
Website: http://www.fema.gov

Friends of the Earth (FOE)

1025 Vermont Avenue NW, Suite 300
Washington, D.C. 20005-6303
Telephone: (202) 783-7400
Fax: (202) 783-0444
E-mail: foe@foe.org
Website: http://www.foe.org

Green Seal

1001 Connecticut Avenue NW, Suite 827
Washington, D.C. 20036-5525
Telephone: (202) 872-6400
Fax: (202) 872-4324
Website: http://www.greenseal.org

Greenpeace USA

1436 U Street NW
Washington, D.C. 20009
Telephone: (202) 462-1177
Website: http://www.greenpeaceusa.org/
Publication: *Greenpeace Magazine*

Hawkwatch International

P.O. Box 660
Salt Lake City, UT 84110
Telephone: (801) 524-8511
E-mail: hawkwatch@charitiesusa.com
Website: http://www.vpp.com/HawkWatch

Humane Society of the United States (HSUS)

2100 L Street NW
Washington, D.C. 20037
Website: http://www.hsus.org
Publications: *All Animals*, a quarterly magazine

International Atomic Energy Commission

P.O. Box 100
Wagramer Strasse 5
A-1400, Vienna, Austria
E-mail: Official.Mail@iaea.org
Website: http://www.iaea.org

International Council for Local Environmental Initiatives (ICLEI)

World Secretariat
16th Floor, West Tower, City Hall
Toronto, M5H 2N2, Canada
Fax: (416) 392-1478
Email: iclei@iclei.org
Website: http://www.iclei.org

International Rhino Foundation (IRF)

14000 International Road
Cumberland Ohio 43732
E-mail: IrhinoF@aol.com
Website: http://www.rhinos-irf.org

International Whaling Commission (IWC)

The Red House
135 Station Road
Impington, Cambridge CB4 9NP,
    United Kingdom
E-mail: iwc@iwcoffice.org
Website: http://ourworld.compuserve.com/
    homepages/iwcoffice

International Wolf Center

1396 Highway 169
Ely, MN 55731-8129

Telephone: (218) 365-4695
Fax: (218) 365-3318
Website: http://www.wolf.org

Jane Goodall Institute (JGI)

P.O. Box 14890
Silver Spring, MD 20911-4890
Telephone: (301) 565-0086
Fax: (301) 565-3188
E-mail: JGIinformation@janegoodall.org

Keep America Beautiful

1010 Washington Boulevard
Stamford, CT 06901
Telephone: (203) 323-8987
Fax: (203) 325-9199
E-mail: info@kab.org

League of Conservation Voters

1707 L Street, NW, Suite 750
Washington, D.C. 20036
Telephone: (202) 785-8683
Fax: (202) 835-0491
E-mail: lcv@lcv.org
Website: http://www.lcv.org

Mountain Lion Foundation (MLF)

P.O. Box 1896
Sacramento, CA 95812
Telephone: (916) 442-2666
E-mail: MLF@moutainlion.org
Website: http://www.mountainlion.org

National Alliance of River, Sound, and Bay Keepers

P.O. Box 130
Garrison, NY 10524
Telephone: (800) 217-4837
E-mail: keepers@keeper.org
Website: http://www.keeper.org

National Anti-Vivisection Society (NAVS)

53 West Jackson Street, Suite 1552
Chicago, IL 60604
Telephone: (800) 888-NAVS
E-mail: navs@navs.org
Website: http://www.navs.org

National Arbor Day Foundation

100 Arbor Avenue
Nebraska City, NE 68410
Telephone: (402) 474-5655
Website: http://www.arborday.org
Publication: *Arbor Day*, a bimonthly magazine

National Audubon Society (NAS)

700 Broadway
New York, NY 10003
Telephone: (212) 979-3000
Website: http://www.audubon.org
Publication: *Audubon*, a bimonthly magazine

National Center for Environmental Health

Mail Stop F-29
4770 Buford Highway NE
Atlanta, GA 30341-3724
Telephone NCEH Health Line: (888)
   232-6789
Website: http://www.cdc.gov/nceh/
   ncehhome.htm

National Parks and Conservation Association (NPCA)

1015 Thirty-first Street NW
Washington, D.C. 20007
Telephone: (202) 944-8530; (800) NAT-PARK
E-mail: npca@npca.org
Website: http://www.npca.org
Publication: *National Parks*, a bimonthly
   magazine

National Wildlife Federation (NWF)

8925 Leesburg Pike
Vienna, VA 22184-0001
Telephone: (800) 822-9919
Website: http://www.nwf.org
Publication: *National Wildlife*, a bimonthly
   magazine

Natural Resources Defense Council (NRDC)

40 West Twentieth Street
New York, NY 10011
Website: http://www.nrdc.org
Publications: *Amiscus Journal*, a quarterly
   magazine

Nature Conservancy (TNC)

1815 North Lynn Street
Arlington, VA 22209
Telephone: (703) 841-5300
Fax: (703) 841-1283
Website: http://www.tnc.org
Publication: *Nature Conservancy*, a magazine

Noise Pollution Clearinghouse

P.O. Box 1137
Montpelier, VT 05601-1137
Telephone: (888) 200-8332
Website: http://www.nonoise.org

North Sea Commission

Business and Development Office
Skottenborg 26, DK-8800 Viborg,
   Denmark
Website: http:\\www.northsea.org

People for Animal Rights

P.O. Box 8707
Kansas City, MO 64114
Telephone: (816) 767-1199
E-mail: parinfo@envirolink.org
Website: http://www.parkc.org

People for the Ethical Treatment of Animals (PETA)

501 Front Street
Norfolk, VA 23510
Telephone: (757) 622-PETA
Fax: (757) 622-0457
Website: http://www.peta-online.org/

Orangutan Foundation International

822 South Wellesley Avenue
Los Angeles, CA 90049
Telephone: (800) ORANGUTAN
Fax: (310) 207-1556
E-mail: ofi@orangutan.org
Website: http://www.ns.net/orangutan

Ozone Action

1700 Connecticut Avenue NW, Third Floor
Washington, D.C. 20009
Telephone: (202) 265-6738

E-mail: cantando@essential.org
Website: www.ozone.org

Peregrine Fund

566 West Flying Hawk Lane
Boise, ID 83709
Telephone: (208) 362-3716
Fax: (208) 362-2376
E-mail: tpf@peregrinefund.org
Website: http://www.peregrinefund.org

Rachel Carson Council

8940 Jones Mill Road
Chevy Chase, MD 20815
Telephone: (301) 652-1877
E-mail: rccouncil@aol.com
Website: http://members.aol.com/rccouncil/
ourpage

Rainforest Action Network

221 Pine Street, Suite 500
San Francisco, CA 94104-2740
Telephone: (415) 398-4404
Fax: (415) 398-2732
E-mail: rainforest@ran.org
Website: http://www.ran.org

Range Watch

45661 Poso Park Drive
Posey, CA 93260
Telephone: (805) 536-8668
E-mail: rangewatch@aol.com
Website: http://www.rangewatch.org

Raptor Resource Project

2580 310th Street
Ridgeway, IA 52165
E-mail: rrp@salamander.com
Website: http://www.salamander.com~rpp

Reef Relief

201 William Street
Key West, FL 33041
Telephone: (305) 294-3100
Fax: (305) 923-9515
E-mail: reef@bellsouth.net
Website: http://www.reefrelief.org

ReefKeeper International

2809 Bird Avenue, Suite 162
Miami, FL 33133
Telephone: (305) 358-4600
Fax: (305) 358-3030
E-mail: reefkeeper@reefkeeper.org
Website: http://www.reefkeeper.org

Renewable Energy Policy Project-Center for Renewable Energy and Sustainable Technology (REPP-CREST)

National Headquarters
1612 K Street, NW, Suite 202
Washington, D.C. 20006
Website: http://www.solstice.crest.org

Resources for the Future (RFF)

1616 P Street NW
Washington, D.C. 20036
Telephone: (202) 328-5000
Fax: (202) 939-3460
E-mail: info@rff.org
Website: http://www.rff.org

Roger Tory Peterson Institute

311 Curtis Street
Jamestown, NY 14701
Telephone: (716) 665-2473
E-mail: webmaster@rtpi.org

Sierra Club

85 Second Street, Second Floor
San Francisco, CA 94105
Telephone: (415) 977-5630
Fax: (415) 977-5799
E-mail (general information):
    information@sierraclub.org
Website: http://www.Sierraclub.org
Publication: *Sierra*, a bimonthly magazine

Smithsonian Institution Conservation & Research Center (CRC)

Website: http://www.si.edu/crc/brochure/
index.htm

Society of American Foresters

5400 Grosvenor Lane
Bethesda, MD 20814

Telephone: (301) 897-8720
Fax: (301) 897-3690
E-mail: safweb@safnet.org
Website: http://www.safnet.org

Surfrider Foundation USA

122 South El Camino Real, Suite 67
San Clemente, CA 92672
Telephone: (949) 492-8170
Fax: (949) 492-8142
Website: http://www.surfrider.org

Union of Concerned Scientists

National Headquarters
2 Brattle Square
Cambridge, MA 02238
Telephone: (617) 547-5552
E-mail: ucs@ucsusa.org
Website: http://www.ucsusa.org
Publications: *Nucleus*, a quarterly magazine;
    Earthwise, a quarterly newsletter

United Nations Environment Programme (Regional)

2 United Nations Plaza
NY, NY 10017
Telephone: (212) 963-8138
Website: http://www.unep.org

United Nations Food and Agriculture Organization (FAO)

Website: http://www.fao.org
Liaison office with North America
Suite 300, 2175 K Street NW, Washington D.C.
    20437-0001

United Nations Man and the Biosphere Programme (UNMAB)

U.S. MAB Secretariat, OES/ETC/MAB
Department of State
Washington, D.C. 20522-4401
Website: http://www.mabnet.org

U.S. Department of Agriculture (USDA)

14th Street and Independence Avenue., SW,
Washington, D.C. 20250
Website: http://www.usda.gov

U.S. Department of Energy (DOE)

Forrestal Building
1000 Independence Avenue, SW,
Washington, D.C. 20585
Website: http://www.doe.gov

U.S. Environmental Protection Agency (EPA)

401 M Street SW
Washington, D.C. 20460
Website: http://www.epa.gov

U.S. Fish and Wildlife Service (FWS)

1849 C Street NW
Washington, D.C. 20240
Telephone: (202) 208-5634
Website: http://www.fws.org

U.S. Geological Survey (USGS)

U.S. Dept. of Interior
1849 C Street, NW
Washington, D.C. 20240
Website: http://www.usgs.gov

U.S. National Park Service (NPS)

U.S. Dept. of Interior
1849 C Street, NW
Washington, D.C. 20240
Website: http://www.nps.gov

U.S. Nuclear Regulatory Commission (NRC)

One White Flint North
11555 Rockville Pike
Rockville, Maryland 20852
Website: http://www.nrc.gov

Wilderness Society

900 Seventeenth Street NW
Washington, D.C. 20006-2506
Telephone: (800) THE-WILD
Website: www.wilderness.org

Wildlands Project (TWP)

1955 West Grant Road, Suite 145
Tucson, AZ 85745
Telephone: (520) 884-0875
Fax: (520) 884-0962

E-mail: information@twp.org
Website: http://www.twp.org

World Conservation Monitoring Centre (WCMC)

219 Huntington Road
Cambridge CB3 ODL, United Kingdom
E-mail: info@wcmc.org.uk
Website: http://www.wcmc.org.uk

World Conservation Union (IUCN)

1630 Connecticut Avenue NW, Third Floor
Washington, D.C. 20009-1053
Telephone: (202) 387-4826
Fax: (202) 387-4823
E-mail: postmaster@iucnus.org
Website: http://www.iucn.org

World Health Organization (WHO)

Avenue Appia 20
1211 Geneva 27
Switzerland
Website: http://www.eho.int
E-mail: inf@who.int

World Parrot Trust United States

P.O. Box 50733
Saint Paul, MN 55150
Telephone: (651) 994-2581
Fax: (651) 994-2580
E-mail: usa@worldparrottrust.org

United Kingdom

Karen Allmann, Administrator,
Glanmor HouseHayle,
Cornwall TR27 4HY,
United Kingdom
E-mail: uk@worldparrottrust.org

Australia

Mike Owen
7 Monteray Street
Mooloolaba, Queensland 4557, Australia
E-mail: australia@worldparrottrust.org
Website: http://www.world parrottrust.org

World Resources Institute

1709 New York Avenue NW
Washington, D.C. 20006
Telephone: (202) 638-6300
E-mail: info@wri.org
Website: http://www.wri.org/wri/biodiv

World Society for the Protection of Animals (WSPA)

P.O. Box 190
Jamaica Plain, MA 02130
Website: http://www.wspa.org

United Kingdom Division
Website: http://www.wspa.org.uk/home.html

World Wildlife Fund, US (WWF)

1250 Twenty-fourth Street NW
P.O. Box 97180
Washington, D.C. 20077-7180
Telephone: (800) CALL-WWF
Website: http://www.worldwildlife.org

WorldWatch Institute

1776 Massachusetts Avenue NW
Washington, D.C. 20036
Telephone: (202) 452-1999
Website: http://www.worldwatch.org/
Publications: *WorldWatch, State of the World, Vital Signs* (annuals)

Zero Population Growth

1400 Sixteenth Street NW, Suite 320
Washington, D.C. 20036
Telephone: (202) 332-2200
Fax: (202) 332-2302
E-mail: zpg@igc.apc.org
Website: http://www.zpg.org

Zoe Foundation

983 River Road
Johns Island, SC 29455
Telephone: (803) 559-4790
E-mail: savage@awod.com
Website: http://www.2zoe.com

INDEX

f indicates figures and photos; t indicates tables

plant, **2:**103, **5:**48–49, **5:**49f
world, **2:**102t
Aquatic ecosystems, succession in,
   1:133–134, **1:**134f
Aquatic life, acid rain effects on, **4:**23f
Aqueduct(s), **2:**91
Aquifer(s), **2:**93, **2:**93f, **4:**43–45, **4:**43f,
   4:45f
   defined, **1:**99, **3:**49, **4:**43f, **4:**55
   depletion of, **4:**44–46, **4:**44t, **4:**45f
   in Great Plains, in America, **3:**41
   Ogallala, **2:**98, **2:**99f, **3:**41, **4:**45, **4:**45f
Arable, defined, **3:**31
Archaebacteria, **1:**19–20, **1:**19f, **1:**20f,
   1:20t
Archeological, defined, **2:**125
Arctic
   air pollution in, **4:**6–7, **4:**6f, **4:**7f
   carbon monoxide in, **4:**6–7, **4:**7f
Arizona's Petrified Forest, **2:**111
Arsenic, **4:**70–71
Art, in Stone Age, **3:**6
Artesian well, **2:**94, **2:**94f
Asbestos, **2:**71
   indoor air pollution by, **4:**12t, **4:**13
Asia, stone tools in, **3:**7
Assembly line, defined, **3:**72
Association of Consulting Foresters, **5:**58
Atmosphere, **1:**7–11, **1:**8f, **1:**9f, **1:**11f,
   4:18–19, **4:**19f
   climate, **1:**10–11, **1:**11f
   gases in, **1:**7–8
   greenhouse effect on, **1:**8, **1:**8f
   layers of, **1:**8–10, **1:**9f
Atom(s), defined, **2:**31
Atomic Energy Commission (AEC), **2:**26
Australia's Great Barrier Reef, **1:**116,
   1:117f
Automobile(s)
   fuel cell, **2:**52–53, **2:**52f, **2:**53f
   fuel cells for, **5:**31–33, **5:**32f, **5:**33f
   during Industrial Revolution, **3:**63–64,
      3:64f
   manufacturing of, eco-efficiency in,
      5:73, **5:**73f
AWEA. *See* American Wind Energy
   Association (AWEA)
Axe(s), hand, **3:**5–6, **3:**5f

Bacteria
   chemosynthetic, **1:**22–23
   coliform, **4:**40, **4:**41
   defined, **5:**51
   diseases caused by, **1:**23t
   round, **1:**23f
   spiral, **1:**23f
Bad breathing cities, **4:**1t
Bagasse, **2:**80, **5:**59
Bald eagle, **5:**68f, **5:**69t
Barter economy, defined, **3:**85
Bat(s), endangered, **4:**115t, **4:**116f
Battery(ies)
   INMETCO Recycling Facility's
      acceptance of, **5:**76t
   recycling of, by INMETCO recycling
      facility, **4:**65t
Bay of Fundy, **1:**115, **1:**115f
Beauty aids, waterway contamination
   due to, **4:**37

Becquerel, Edmond, **2:**42, **5:**25
Beef, as food source, **3:**95
Bench terracing, **5:**40, **5:**40f
Benthic zone, **1:**104
   defined, **1:**121
   of open ocean zone, **1:**119–120, **1:**121f
Benzene, **4:**72–23
Bering Sea, **2:**101, **2:**101f, **5:**35f
Bessemer process, defined, **3:**72
Beverage Container Act (Bottle Bill),
   4:66–67
Bike commuting, **5:**99, **5:**100f
Biocentrism, **5:**14
Biodegradable, defined, **4:**69
Biodiversity, **1:**65–66, **4:**109–111, **4:**111t
   defined, **1:**77, **4:**109, **5:**15
   protecting of, **5:**3, **5:**60–64, **5:**62t,
      5:63f, **5:**63t
   treaties, laws, and lists in, **4:**111–116,
      4:113f–116f, **4:**113t, **4:**115f
   value of, **4:**110–111
Biodiversity Treaty, **4:**111, **5:**60
Biofuel(s), **2:**48–49, **5:**28
   environmental concerns of, **2:**49
Biogeochemical cycles, **4:**18
   defined, **4:**34
Biological control, **5:**41
   defined, **5:**51
Biological diversity, changes in,
   deforestation and, **4:**93–94
Biomass, **2:**47–49, **5:**27–28
   defined, **1:**54
   described, **2:**47–48, **5:**27–28
Biome(s(s)), **1:**18, **1:**18f
   average annual rainfall of, **1:**58f, **1:**70f,
      1:81f, **1:**92f, **1:**97f
   defined, **2:**72, **4:**101
   freshwater. *See* Freshwater biomes
   land, **1:**56–78. *See also* Forest(s)
   marine, **1:**109–122. *See also* Marine
      biomes
   water, **1:**101–122. *See also* Water biomes
Biosphere, **1:**2–3, **1:**2f, **1:**17–39
   components of, **1:**2
   defined, **1:**2
   described, **1:**2, **1:**2f
   size of, **1:**2
Biosphere Reserves, **5:**65
Biotechnology, described, **5:**44–46, **5:**45t
Biphenyl(s), polychlorinated, **4:**73
Bird(s), **1:**27
   dodo, **4:**103, **4:**104f
Birthrate, measuring of, **3:**88
Bison
   in Great Plains, **3:**42f
   in Yellowstone National Park, **2:**115f
Black Death, defined, **3:**100
"The Black Land," **3:**18
Blue Angel, **5:**78
Bog(s), **1:**14, **1:**106, **1:**107f
Boiling water reactor (BWR), **2:**25
Bone tools, in New Stone Age, **3:**7f, **3:**8,
   3:8f
Bottle Bill, **4:**64–66
Bottle gas, **2:**7
Bowhead, **1:**119f
Breaker boys, **3:**58
Breeding
   livestock, **3:**37
   selective, defined, **3:**49
Brick(s), in Mesopotamia, **3:**16
British Thermal Units (BTUs), **2:**12
Bronze Age, **3:**15
Brower, David, **5:**13
Brownfield(s), **4:**79, **4:**79f, **4:**80f
   cleaning up of, **5:**89–90, **5:**89–91f

defined, **5:**102
Swedish housing on, **5:**91, **5:**91f
Brundtland Report, **5:**9–10
BTUs. *See* British Thermal Units
   (BTUs)
Bullard, Linda McKeever, **5:**109
Bullard, Robert, **5:**109, **5:**110f
Bullitt Foundation, **5:**113, **5:**113f
Bureau of Land Management (BLM),
   2:109, **2:**119–121, **2:**119f–121f
Bus(es), **5:**96, **5:**97f
   cleaner fuels for, **5:**96–97, **5:**97f, **5:**98f
   electric, **5:**97f
   propane, **5:**97f
Business stewardship
   eco-efficiency in, **3:**85, **3:**85f, **5:**71–72.
      See also Eco-efficiency
   eco-labeling in, **5:**78–79
   sustainable, **5:**71–84. *See also* Sustainable
      business stewardship
BWR. *See* Boiling water reactor (BWR)

CAA. *See* Clean Air Act (CAA)
Clean Air Act, **1:**9
Cactus(i), **1:**91–92, **1:**93f, **1:**94
   endangered, **4:**113f, **4:**113t
Cadmium, **4:**73
   defined, **2:**31
Calcium cycle, **1:**54
California condor, **4:**119f, **4:**120t
Canada, stone tools in, **3:**7
Canada Deuterium Uranium (Candu)
   reactor, **2:**25
Candu reactor, **2:**25
Canopy, defined, **1:**77
Cape Hatteras National Seashore, **2:**111
Capital, defined, **3:**85
Capital resources
   defined, **5:**83
   production of, **3:**76–77
Capitalism, defined, **3:**81, **3:**85
Captive propagation, **4:**117–120,
   4:118f–120f, **4:**118t
   defined, **5:**69
Captive propagation programs, **5:**65
Carara Biological Reserve, **5:**66, **5:**67
Carbohydrate, defined, **1:**54
Carbon, in environment, **1:**50–51
Carbon cycle, **1:**50–52, **1:**51f, **4:**27
Carbon dioxide, **4:**26–27, **4:**27f
   atmospheric concentrations of, **5:**11f
   defined, **5:**33
   global warming due to, **5:**58
   sources of, **4:**29f, **5:**17, **5:**17f
Carbon dioxide molecule, **4:**7f
Carbon monoxide, in Arctic, **4:**6–7, **4:**7f
Carbon monoxide molecule, **4:**7f
Carcinogen(s), defined, **4:**87
Caribbean Plate, **1:**4, **1:**5f
"Caring for the Land and Serving
   People," **2:**115
Carnivore(s), in ecosystem, **1:**42–43, **1:**43f
Carrying capacity
   defined, **3:**100, **4:**101, **5:**15
   of population, **3:**90–91
Carson, Rachel Louise, **5:**114–115
Cartwright, Edmund, in Industrial
   Revolution in America, **3:**53–54
Catalyst
   defined, **4:**16, **5:**33
   described, **5:**30
Catalytic converters, in air pollution
   control, **4:**10
Cat's Claw, **2:**83, **2:**83f
Cell(s)
   defined, **2:**105

Darwin's theory of evolution, **1:**135
dBA. *See* Decibel (dBA)
DDT, defined, **5:**117
Death, Black, defined, **3:**100
Death rate, measuring of, **3:**88–89
Decibel (dBA), **4:**15
  defined, **4:**16
Deciduous, defined, **1:**77
Deciduous forest(s), temperate, **1:**62–66,
  1:63f–67f
Deciduous forest animals, **1:**65
Deciduous forest plants, **1:**63–65
Decomposer(s), **1:**43–44, **1:**44f
Decomposition, aerobic, defined, **4:**69
Deep ecology, **5:**14
Deep zone, of open ocean zone,
  1:119–120, **1:**121f
Deepwell injection, in hazardous waste
  disposal, **4:**79–80
Deforestation, **4:**89–94
  biological diversity changes due to,
    4:93
  clear-cutting and, **4:**92–93, **4:**92f
  consequences of, **4:**93–94
  controlling of, **4:**101
  defined, **2:**87, **3:**18, **3:**31, **5:**54, **5:**69
  global environmental changes due to,
    4:94
  problems related to, solving of, **4:**94
  regulation of, **4:**101
  slash-and-burn practices and,
    4:90–92, **4:**91f
  soil quality changes due to, **4:**93–94
Degradation, land, causes of, **5:**3f
Demographer(s), defined, **3:**100
Denmark, wind plants of, **2:**35–36
Deoxyribonucleic acid (DNA)
  defined, **1:**54
  nitrogen in, **1:**52
Department of Energy (DOE), **2:**29f
Department of Health and Human
  Services (DHHS), **2:**9
Department of Housing and Urban
  Development, **2:**68
Desalination, **2:**104–105, **2:**104f
Desert(s), **1:**88–92, **1:**90f–93f
  animals in, **1:**89–92, **1:**93f
  average annual rainfall of, **1:**92f
  described, **1:**88–89
  in Eastern and Western hemispheres,
    1:90f
  environmental concerns of, **1:**92, **1:**93
  life in, **1:**89–92, **1:**93f
  North American, **1:**91f
Desert soils, **2:**62
Desertification, **4:**96–97, **4:**96f–97f
Detritus, defined, **2:**72
Deuterium
  defined, **2:**31
  in nuclear fusion, **2:**29–30, **2:**30f
Developed countries, defined, **3:**100
Developing countries
  defined, **2:**87, **3:**100, **5:**15
  population growth in, **3:**89
DHHS. *See* Department of Health and
  Human Services
Diesel fuel, **2:**7
Diet(s), in Stone Age, **3:**10f
Digging tools, early, **3:**33f

Disease(s), human health and, **3:**98
Disfranchise, defined, **5:**117
Dissolved oxygen, defined, **4:**55
Distillation, fractional, **2:**6
  defined, **2:**18
Diversity, defined, **2:**72
DNA. *See* Deoxyribonucleic acid
  (DNA)
Dodo bird, **1:**137, **1:**138f, **4:**103, **4:**104f
DOE. *See* Department of Energy (DOE)
Domesticated plants, **3:**9–10, **3:**9t
Domestication, defined, **3:**11
Douglas, Marjory Stoneman, **5:**114
Dove, Nevada, **5:**108, **5:**108f
Downy mildew, **1:**24
Draft Declaration, **5:**6
Drake, Edwin, **2:**3
Drinking water
  treating of, in water pollution
    treatment, **4:**47
  treatment of, for safety, **2:**91–92, **2:**92t
Drip irrigation, **2:**96–97, **2:**98f,
  5:42–43, **5:**43f
Dumping in Dixie, **5:**109
DuPont Corporation, **5:**71
Dust Bowl, **3:**48
  defined, **3:**49
Dust Bowl region, **3:**48f

E85, **5:**28
Eagle(s), bald, **5:**68f, **5:**69t
Early ancient tools, **3:**3f
Early human civilizations, **3:**13–32
Early stone tool communities, **3:**2, **3:**2f
Earth
  surface of
    broken up nature of, **1:**3–5
    photograph of, **1:**1, **1:**1f
  systems of, **1:**1–16
Earth Day, **5:**112–113, **5:**113f
Earth Policy Institute, **5:**20
Earth Summit, **5:**10, **5:**11, **5:**60, **5:**80,
  5:107
Earth systems, **1:**1–16. *See also*
  Lithosphere; specific types, e.g.,
  Biosphere
Earthquake(s)
  ecosystem effects of, **1:**124–125,
    1:125f, **1:**127t
  magnitudes of, **1:**127t
Earth's crust, **1:**3
Earth's troposphere, **4:**18–19
Eastern states, land cleared by, **3:**47f
Eastside Community Garden and
  Education Center, **5:**112
Eco-economy, professions in, **5:**83t
Eco-efficiency, **3:**84–85, **3:**85f, **5:**71–84
  accounting for, **5:**80–83, **5:**80t,
    5:81f–83f, **5:**83t
  goals of, **5:**71–72
  life cycle of product and, **5:**72–77,
    5:73f, **5:**75f, **5:**76f
  in manufacturing automobiles, **5:**73,
    5:73f
  in schools, **5:**77
Eco-efficiency companies, **5:**74–76,
  5:75f, **5:**76t
Eco-industrial parks, **5:**77
Eco-label(s)
  caution related to, **5:**79
  examples of, **5:**78
Eco-labeling, **5:**78–79
  of fish, **5:**50, **5:**50f
Ecological pyramids, **1:**48–49, **1:**48f
Ecological succession, **1:**130–134,
  1:131f, **1:**133f, **1:**134f

in aquatic ecosystems, **1:**133–134,
  1:134f
defined, **4:**101
pioneer species, **1:**132–133, **1:**133f
primary succession, **1:**130–131, **1:**131f
secondary succession, **1:**131–132
Ecologists, defined, **1:**38
Economic(s)
  defined, **3:**86, **5:**15
  environmental impact of, **3:**84
Economic expansion
  environmental impact of, **3:**73–86
  goods, **3:**74–77, **3:**75t, **3:**76f
  models of, **3:**81, **3:**81f
  services, **3:**74–77, **3:**75t, **3:**76f
Economic growth, defined, **3:**86, **5:**83
Economist(s), defined, **3:**100
Economy
  of ancient China, **3:**29
  of ancient Egypt, **3:**20
  of ancient Greece, **3:**23
  of ancient Rome, **3:**26, **3:**25f
  barter, defined, **3:**85
  defined, **3:**86
  measuring of, **3:**81–84, **3:**82t, **3:**83f
  of Mesopotamia, **3:**16–17
  sustainable
    creation of, **5:**7–8, **5:**8t
  teen-age projects on, **5:**81–82, **5:**82f
Ecosystem(s), **1:**17–18, **1:**17f, **1:**40–55
  acid rain effects on, **4:**22–23, **4:**23f
  aquatic, succession in, **1:**133–134,
    1:134f
  carnivores in, **1:**42–43, **1:**43f
  changes in, **1:**123–141
  consumers in, **1:**42–43, **1:**43f
  decomposers in, **1:**43–44, **1:**44f
  defined, **3:**31, **4:**121
  energy flows through, **1:**44–49, **1:**45f,
    1:47f, **1:**48f
  evolution effects on, **1:**134–135
  extinction effects on,
    1:136–139, **1:**136f–138f
  herbivores in, **1:**42, **1:**43f
  human effects on, **1:**139–140
  living organisms in, **1:**18–29, **1:**19f,
    1:20f, **1:**20t, **1:**21f–23f, **1:**23t,
    1:25f–29f, **1:**40–44, **1:**41f–44f
  matter flows through, **1:**49–54, **1:**50f,
    1:51f, **1:**53f
  natural disasters effects on, **1:**123–130,
    1:123f–126f, **1:**127t, **1:**128f, **1:**129f
  omnivores in, **1:**43
  phytoplankton in, **1:**41, **1:**42f
  producers in, **1:**40–42, **1:**41f, **1:**42f
  rain forest, **1:**70
  scavengers in, **1:**42–43, **1:**43f
  taiga forest, **1:**59–60
  temperate rain forest, **1:**67–68, **1:**67f
  threats to, **4:**44t
Ecoteams, **5:**93
Ecotone, defined, **5:**69
Ecotourism, defined, **1:**99, **5:**69
Ectothermic, defined, **1:**27, **1:**38
Egypt, ancient, **3:**14f, **3:**18–21, **3:**19f
  agriculture in, **3:**19, **3:**19f
  economy of, **3:**20
  geography of, **3:**18
  human impact on, **3:**20–21
  irrigation in, **3:**18–19
  natural resources of, **3:**18
  soil resources in, **3:**18, **3:**19f
  trade of, **3:**20
  water resources in, **3:**18–19
Egyptian(s), early, gifts and benefits of,
  3:20

Feldspar, **2:**65
Fertile Crescent, **3:**15, **3:**15f
 legacy of, **3:**17
Fertilizer(s)
 history of, **3:**36
 natural, **3:**35
 pollution due to, **4:**39
Fertilizer runoff, agricultural pollution
  due to, **4:**95
FIFRA. *See* Federal Insecticide,
  Fungicide, and Rodenticide Act
  (FIFRA)
Finch species, on Galapagos Islands,
  1:136f
Finland, stone tools in, **3:**7
Fire(s), in Stone Age, **3:**3–4
Firewood, **5:**28
Fish, **1:**29, **1:**29f
 eco-labeling of, **5:**50, **5:**50f
Fish and Wildlife Service (FWS), **2:**109,
  4:112–113, **4:**117, **5:**61, **5:**64, **5:**65
Fish and Wildlife Service (FWS)
  Endangered Species, **5:**63
Fish aquaculture, **2:**100–102,
  2:101f–102f
Fish stocks, rebuilding of, **5:**49–50
Fishing
 commercial, **2:**100t, **5:**46, **5:**46t
 oceanic
  environmental concerns of, **2:**99–100
  as food source, **3:**95
 sustainable, **5:**46–51, **5:**46f, **5:**46t, **5:**49f,
   5:50f. *See also* Sustainable fishing
Fission, described, **2:**22f
Flint, **3:**3
Flint tools, **3:**3–4
Flood(s), ecosystem effects of, **1:**124
Flood irrigation, **2:**95–96
Floodplain, defined, **1:**121, **3:**32
Florida Everglades, **5:**114
Florida Everglades National Park,
  defined, **5:**117
Florida panther, **5:**62t, **5:**63f
Florida's Pelican Island, **4:**117
Fluidized bed combustion, reducing of,
  4:23–24
Fluoride, defined, **2:**105
Foliage, defined, **1:**140
Food
 electricity and, during Industrial
  Revolution, **3:**66, **3:**67f
 from forests, **2:**80–81
 production and distribution of, **3:**94–98
 sources of, **3:**94–96
  crops, **3:**94–95
  livestock, **3:**95
  oceanic fishing and aquaculture,
   3:95–96
Food and Agriculture Organization
  (FAO), of U.N., **5:**45–46, **5:**47, **5:**54
Food and Drug Administration (FDA),
  4:41
Food chain, **1:**44, **1:**45f
Food distribution, electricity and,
  during Industrial Revolution, **3:**66,
  3:67f
Food sources
 in Mesopotamia, **3:**15
 from ocean, **2:**98, **2:**99f

Food webs, **1:**46–46, **1:**47f
Ford, **5:**33
Ford, Henry, **3:**83, **5:**73f
Forest(s), **1:**56–78, **2:**74–88
 acid rain effects on, **4:**22
 benefits of, **5:**54–56
 contributions of, **2:**75–86
 described, **2:**74
 disappearing, **3:**47f, **5:**54–56, **5:**54f,
   5:55f
 Eastern and Western hemispheres, **1:**57f
 environmental concerns of, **2:**86–87
 fuelwood from, **2:**75–76, **2:**75f, **2:**76t
 land surface covered by, **5:**53
 loss of, **4:**89–102, **5:**54. *See also*
   Deforestation
  prospect of, **5:**53
 management of, **2:**81
 medicinal products from, **2:**81–86,
   2:82f-85f
 non-timber products from, **2:**80–81
 old growth, defined, **2:**87
 old-growth, **1:**75
 paper products from, **2:**78–79, **2:**79t,
   2:80f
 rainforests, **2:**83–86, **2:**83f–85f. *See*
   also Rainforest(s)
 temperate, **1:**66–67. *See also* Temperate
   rain forests
 sustainability of, management of,
   5:56–59, **5:**57f, **5:**59f
 sustainable, **5:**53–59. *See also* Sustainable
   forests
 taiga, **1:**56–61, **1:**56f–61f. *See also* Taiga
   forests
 temperate, **1:**61–68, **1:**62f–65f, **1:**67f.
   See also Temperate forests
  in Eastern and Western hemispheres,
   2:74f
  environmental concerns of, **1:**65–66
 tropical rain, **1:**68–75, **1:**69f–74f
 urban. *See* Urban forests
  in U.S., **2:**75f
  prevalence of, **5:**54
 wood products from, **2:**77–78, **2:**78t
Forest fires, ecosystem effects of, **1:**124,
  1:124f
Forest Service in 1905, **2:**115
Forest Stewardship Council (FSC), **5:**58
Fossil fuel(s), **2:**1–18. *See also specific types*
 coal resources, **2:**11–17, **2:**13f–15f,
   2:17t
 described, **2:**1
 gasoline, **2:**6–9
 natural gas, **2:**9–11, **2:**10t, **2:**11f
 petroleum, **2:**3–6, **2:**4f–6f
 resources from, **2:**1–9, **2:**2t, **2:**3f–6f,
   2:8t
Fossil fuel emissions, environmental
  effects of, **5:**17–18
Fractional distillation, **2:**6
 defined, **2:**18
Franklin, B., as paper merchant, **2:**79
Free enterprise system, **3:**80–81, **3:**81f
Fresh Kills landfill, **4:**62
Freshwater
 sources of, **2:**91–94, **2:**92, **2:**93f, **2:**94f,
   4:37f
 uses of, **2:**89–98, **2:**90f, **2:**91f, **2:**92t,
   2:93f, **2:**95t, **2:**97f, **2:**97t, **2:**98f
Freshwater biomes, **1:1:**101f, **1:**101,
  1:102–109, **1:**103t, **1:**104f–107f
 bogs, **1:**106, **1:**107f
 freshwater swamps, **1:**105–106
 lakes, **1:**104, **1:**104f
 marshes, **1:**107–108

 ponds, **1:**103
 rivers, **1:**102–103, **1:**103t
 wetlands, **1:**105–108, **1:**107f
Freshwater bodies, **1:**13–14, **1:**14f
Freshwater marshes, **1:**14
Freshwater pollution, **4:**36–40, **4:**37f,
  4:37t, **4:**38f, **4:**39t
 agricultural pollution—related, **4:**37–40
 potable water—related, **4:**36–37
 thermal pollution—related, **4:**40
Freshwater swamps, **1:**105–106
Friends of McKinley, Inc., **5:**108f
FSC. *See* Forest Stewardship Council
  (FSC)
FTC. *See* Federal Trade Commission
  (FTC)
Fuel(s)
 cleaner, for buses, **5:**96–97, **5:**97f,
   5:98f
 fossil. *See* Fossil fuels; *specific types*
 nuclear, **2:**23
 soybean, **5:**27f
 spent, **2:**28
Fuel cell(s)
 for automobiles, **5:**31–33, **5:**32f, **5:**33f
 electrolytes in, **5:**30
 history of, **5:**33
 sources of, **5:**31f
Fuel cell automobiles, **2:**52–53, **2:**52f,
  2:53f
Fuel Cell Hybrid Vehicle (FCHV),
  Toyota's, **2:**52f
Fuel rods
 defined, **4:**87
 in nuclear reactors, **2:**23–24
Fuelwood, **2:**48
 countries with, **2:**76, **2:**76t
 environmental concerns of, **2:**48, **2:**76
 from forests, **2:**75–76, **2:**75f, **2:**76t
 historical background of, **2:**75–76
 during Industrial Revolution, **3:**56–57
Fungus(i), **1:**20t, **1:**25–26, **1:**25f
 biodiversity of, **5:**4t
Furrow irrigation, **2:**96
Fusion
 described, **2:**19, **2:**30, **2:**30f
 nuclear, **2:**29–30, **2:**30f
FWS. *See* Fish and Wildlife Service
  (FWS)

Gaia, described, **1:**2–3
Gaia hypothesis, **1:**2–3
Garbage, **4:**57
Garbage disposal, in Curitiba, Brazil,
  4:66
The Garden Project, **5:**105f
Gas(es)
 atmospheric, **1:**7–8
 coal, during Industrial Revolution, **3:**59
 greenhouse, **4:**26–27, **4:**27f
 natural
  during Industrial Revolution, **3:**59,
   3:60f
  liquefied. *See* Liquefied natural gas
   (LNG)
 soil, **2:**59–60
Gasoline, **2:**6–9
 described, **2:**6–7
Gateway National Recreation Area,
  2:112
GDP. *See* Gross domestic product (GDP)
Gems, valuable, **2:**69
General Motors, **5:**31, **5:**33, **5:**71
Genetic(s), defined, **3:**49, **4:**121
Geologists, defined, **1:**15
Geothermal, defined, **2:**125

Hydrogen fuel cells, **2:**51–53, **2:**52f, **2:**53f, **5:**30–31, **5:**31f
concerns related to, **2:**53
described, **2:**51–52
source of, **2:**52f
Hydrological cycle, **1:**49f, **1:**50
Hydropower. *See* Hydroelectric power
Hydrosphere, **1:**11–15, **1:**12f–14f
defined, **1:**11
freshwater bodies, **1:**13–14
ocean currents, **1:**12–13, **1:**13f
oceans, **1:**11–12, **1:**12f
wetlands, **1:**14–15
Hypothermia, defined, **4:**55
Hypothesized, defined, **1:**15

Ice age, defined, **1:**99
IFQs. *See* Individual fishing quotients (IFQs)
Ignitable, defined, **4:**71t
Immigration, defined, **3:**100
In situ, defined, **2:**72
Incineration, **4:**61–62
in hazardous waste disposal, **4:**77–78, **4:**78f
Incinerator(s), **4:**62–63, **4:**62f
Income inequality, by country, **3:**99t, **5:**5t
India
ancient, **3:**14f, **3:**30–31, **3:**30f
climate of, **3:**30, **3:**30f
early agriculture in, **3:**31
early legacy of, **3:**31
geography of, **3:**30
monsoons in, **3:**30, **3:**30f
natural resources of, human impact on, **3:**31
stone tools in, **3:**6
tigers in, **4:**103
wind plants of, **2:**36
Indian-Australian Plate, **1:**4, **1:**5f
Indigenous, defined, **2:**87, **3:**11, **5:**15
Indigenous peoples, **3:**3, **3:**4f
human rights for, **5:**6, **5:**6f
of rainforest, **2:**86
Individual fishing quotients (IFOs), **5:**49
Individual transferable quotas (ITQs), **5:**50
Indoor pollutants. *See* Air pollutants, indoor
Industrial Revolution, **3:**50–72
accidents in mines and factories during, **3:**68–69
achievements during, **3:**69
in America, **3:**53–56, **3:**54f, **3:**55f
Edmund Cartwright in, **3:**53–54
factory system, **3:**54–56, **3:**55f
Samuel Slater in, **3:**53, **3:**54f
benefits of, **3:**69
defined, **2:**18
electricity during, **3:**65–66, **3:**67f
communications, **3:**65–66
food and food distribution, **3:**66, **3:**67f
light and power, **3:**65
newspapers, **3:**66
energy source during
changes in, **3:**56–61, **3:**57f, **3:**58f, **3:**60f, **3:**61f
charcoal, **3:**56–57
coal, **3:**56–57, **3:**58f
coal gas, **3:**59

fuel wood, **3:**56–57
iron, **3:**57, **3:**59
natural gas, **3:**59, **3:**60f
steel, **3:**57, **3:**59
waterpower, **3:**56, **3:**57f
wind power, **3:**56
in England, **3:**51–53, **3:**53t
birth of, **3:**52
factory system, **3:**51–52
petroleum resources, **3:**59–60, **3:**61f
fossil fuel demands during, **2:**1
health and medicine in, **3:**67–69
health conditions during, **3:**68
social, cultural, and environmental impact of, **3:**70
timeline of (1701–1909), **3:**70–71
transportation during, **3:**61–65, **3:**62f, **3:**64f
automobiles, **3:**63–64, **3:**64f
new roads, **3:**64–65
railroads, **3:**63
steam engines, **3:**61–62
steam locomotives, **3:**62, **3:**62f
steam power, **3:**61–62
steamships, **3:**62
Industrial smog, **4:**7–8
Inequality, income, by country, **5:**5t
Influenza virus, epidemics of, population decline due to, **3:**92–93, **3:**92f
Infrastructure, **5:**102
defined, **2:**53
INMETCO Recycling Facility, batteries accepted by, **4:**65t, **5:**76t
Inorganic, defined, **2:**72
Inorganic material, defined, **4:**69
Insectivore(s), defined, **1:**38
Integrated pest management (IPM), **5:**40–41, **5:**41f, **5:**42f
Interface Flooring Systems, **5:**76
International Agency for Research on Cancer, **2:**9
International Engine of the Year Awards, **5:**32
International reserves, **4:**117, **5:**65–67, **5:**67f, **5:**68f, **5:**69t
Intertidal zones, **1:**113–114
life in, **1:**113–114
Invertebrate(s), **1:**27, **1:**27f
Ion(s), defined, **2:**73
IPM. *See* Integrated pest management (IPM)
Ireland, stone tools in, **3:**7
Iron
during Industrial Revolution, **3:**57, **3:**59
in Mesopotamia, **3:**16
Irrigated area
in countries (1994), **2:**97t
in United States, top (1997), **2:**95t
Irrigation
in ancient Egypt, **3:**18–19
drip, **2:**96–97, **2:**98f, **5:**42–43, **5:**43f
environmental concerns of, **2:**97
flood, **2:**95–96
furrow, **2:**96
methods of, **2:**95–97, **2:**97f
pivot, **2:**97f, **5:**43f
with saltwater, **5:**44
water for, **2:**95–98, **2:**95t, **2:**97f, **2:**97t, **2:**98f
Isotope(s), defined, **2:**31
Itaipú Hydroelectric Power Plant, **2:**38f, **5:**29f
ITQs. *See* Individual transferable quotas (ITQs)
IUCN. *See* World Conservation Union (IUCN)

Jackson, Simon, **5:**109
John Deere plow, **3:**42, **3:**43f
J-shaped curve, population-related, **3:**91–92, **3:**91f
Julian, Percy, **5:**45f

Kayapo, **2:**86
Kenaf, **5:**59, **5:**59f
Kerosene, **2:**7
Keystone species, **1:**30
Kilowatt (kW), defined, **2:**53
Kimberly Clark papermill, **5:**59
Kinetic energy, **2:**34
defined, **2:**53, **5:**33
Kingdom(s), classification of, **1:**19, **1:**20t
Kudzu, **4:**107
Kyoto Protocol, **4:**34, **5:**10, **5:**12f

Lake(s), **1:**13–14, **1:**104, **1:**104f
environmental concerns of, **1:**105
Lake Baikal, **1:**106, **1:**106f
Land biomes, **1:**56–78. *See also* Forest(s)
Land degradation, causes of, **5:**3f
Land enclosure, **3:**35
Land Ordinance of 1785, **2:**119
Land resources, **2:**56–73
forests, **2:**74–88
minerals, **2:**64–72, **2:**67t, **2:**70f, **2:**71f
phosphorus, **2:**69–70, **2:**70f
soil, **2:**56–64, **2:**57f–59f, **2:**61f, **2:**63f, **2:**64f
Landfill(s), **4:**59–62, **4:**61f, **4:**62f
described, **4:**59–60
design of, **4:**60–62, **4:**61f
Fresh Kills, **4:**62
hazardous wastes in, **4:**72t
PCB, **5:**111
prevalence of, **4:**62
sanitary, **4:**59–62, **4:**61f, **4:**62f
secured, in hazardous waste disposal, **4:**77
Landsat satellite, photograph of Earth's surface by, **1:**1, **1:**1f
Landslide, defined, **1:**140
Larderello geothermal field, **2:**45, **5:**27
Latitude, **1:**10, **1:**10f
Lava, **1:**125
defined, **1:**140
Leach, defined, **4:**34
Leachate, defined, **4:**87
Leaching, defined, **4:**34
Lead, **2:**66–68
Leaf(ves)
maple, **1:**63f
oak, **1:**63f
Lee, Charles, **5:**111
Legume(s), defined, **5:**52
Less-developed countries, population growth in, **3:**89
Levee, defined, **3:**32
Lichen(s), **1:**32.1.**1:**33f, **2:**57f
defined, **1:**77, **4:**16
types of, **4:**9f
Life cycle, of product, eco-efficiency and, **5:**72–77, **5:**73f, **5:**75f, **5:**76f
Light, electricity and, during Industrial Revolution, **3:**65
Light rail transit (LRT), **5:**97–98
Light water reactor (LWR), **2:**24–25
Lignite, **2:**12
Limestone, in ancient Egypt, **3:**20
Liming, **4:**24f
Lion(s), in Serengeti National Park, **5:**67f
Liquefied natural gas (LNG), **5:**97
defined, **5:**102
Liquefied petroleum gas (LPG), **2:**7
Liquid, volatile, defined, **4:**101

ABOUT THE AUTHORS

JOHN MONGILLO is a noted science writer and educator. He is coauthor of *Encyclopedia of Environmental Science*, and *Environmental Activists*, both available from Greenwood.

PETER MONGILLO has won several awards for his teaching, including School District Teacher of the Year, National Endowment for the Humanities Fellowship Award, and the National Council for Geographic Education Distinguished Teacher Award.